ESP8266 Internet of Things Cookbook

Over 50 recipes to help you master the ESP8266's functionality

Marco Schwartz

BIRMINGHAM - MUMBAI

ESP8266 Internet of Things Cookbook

First published: April 2017

Production reference: 1240417

Published by Packt Publishing Ltd.
Livery Place
35 Livery Street
Birmingham B3 2PB, UK.

ISBN 978-1-78728-810-2

www.packtpub.com

Credits

Author
Marco Schwartz

Reviewer
Catalin Batrinu

Acquisition Editor
Prachi Bisht

Content Development Editor
Trusha Shriyan

Technical Editor
Varsha Shivhare

Copy Editor
Safis Editing

Project Coordinator
Kinjal Bari

Proofreader
Safis Editing

Indexer
Francy Puthiry

Graphics
Kirk D'Penha

Production Coordinator
Nilesh Mohite

Cover Work
Nilesh Mohite

About the Author

Marco Schwartz is an electrical engineer, entrepreneur, and blogger. He has a master's degree in electrical engineering and computer science from Supélec, France, and a master's degree in micro engineering from the **Ecole Polytechnique Fédérale de Lausanne** (**EPFL**), Switzerland.

He has more than five years of experience working in the domain of electrical engineering. Marco's interests center around electronics, home automation, the Arduino and Raspberry Pi platforms, open source hardware projects, and 3D printing. He has several websites about the Arduino, including the open home automation website, which is dedicated to building home automation systems using open source hardware. Marco has written another book on home automation and the Arduino, called *Arduino Home Automation Projects*. He has also written a book on how to build Internet of Things projects with the Arduino, called *Internet of Things with the Arduino Yun*, by Packt Publishing.

About the Reviewer

Catalin Batrinu graduated from the Politehnica University of Bucharest in Electronics, Telecommunications, and Information Technology. He has worked as a software developer in telecommunications for the past 16 years.

He started working with old protocols to the latest network protocols and technologies so he caught all the transformations in telecommunication industry.

He has implemented many telecommunication protocols, from access adaptations and backbone switches to high capacity carrier-grade switches on various hardware platforms, such as Wintegra and Broadcom.

Internet of Things came as a natural evolution for him, and now he collaborates with different companies to construct the world of the future that will make our life more comfortable and secure.

Using ESP8266, he has prototyped devices such as irrigation controller, smart sockets, window shutters, lighting control using **Digital Addressable Lighting Control** (**DALC**), and environment control, all of them being controlled directly from a mobile application over the cloud. Even an MQTT broker with a bridging and web socket server has been developed for the ESP8266. Soon, all these devices will be part of our daily life, so we will all enjoy their functionality.

You can read his blog at http://myesp8266.blogspot.com.

www.PacktPub.com

eBooks, discount offers, and more

Did you know that Packt offers eBook versions of every book published, with PDF and ePub files available? You can upgrade to the eBook version at www.PacktPub.com and as a print book customer, you are entitled to a discount on the eBook copy. Get in touch with us at customercare@packtpub.com for more details.

At www.PacktPub.com, you can also read a collection of free technical articles, sign up for a range of free newsletters and receive exclusive discounts and offers on Packt books and eBooks.

https://www.packtpub.com/mapt

Get the most in-demand software skills with Mapt. Mapt gives you full access to all Packt books and video courses, as well as industry-leading tools to help you plan your personal development and advance your career.

Why subscribe?

- ► Fully searchable across every book published by Packt
- ► Copy and paste, print, and bookmark content
- ► On demand and accessible via a web browser

Customer Feedback

Thanks for purchasing this Packt book. At Packt, quality is at the heart of our editorial process. To help us improve, please leave us an honest review on this book's Amazon page at `https://www.amazon.com/dp/1787288102`.

If you'd like to join our team of regular reviewers, you can e-mail us at `customerreviews@packtpub.com`. We award our regular reviewers with free eBooks and videos in exchange for their valuable feedback. Help us be relentless in improving our products!

Table of Contents

Preface

The ESP8266 chip is a powerful and cheap microcontroller with an onboard Wi-Fi connection. It is also very easy to use, thanks to the compatibility with the Arduino IDE. Therefore, it's just the perfect chip to build the Internet of Things (IoT) projects.

Inside this book, we'll see how to build IoT projects using the ESP8266, via several step-by-step tutorials. At the end, you will know how to use all the functions of the ESP8266, and you will be able to build your own projects with this amazing Wi-Fi chip.

What this book covers

Chapter 1, *Configuring the ESP8266*, will be about getting started with the ESP8266, and learning how to configure the ESP8266 and all the hardware/software components that you need to use it.

Chapter 2, *Your First ESP8266 Projects*, will be about learning how to make your first simple projects with the ESP8266.

Chapter 3, *More ESP8266 Functions*, will be about learning advanced functions of the ESP8266, as using the file storage system.

Chapter 4, *Using MicroPython on the ESP8266*, will be focused on using the powerful & simple MicroPython language to build projects with the ESP8266.

Chapter 5, *Cloud Data Monitoring*, will be about connecting your ESP8266 to the cloud, in order to monitor your projects from anywhere in the world.

Chapter 6, *Interacting with Web Services*, will exploit the Wi-Fi connectivity of the ESP8266 to connect to existing Web services, such as IFTTT or Google Drive.

Chapter 7, *Machine to Machine Interactions*, will be about building projects where ESP8266 boards communicate directly with each other via the cloud.

What you need for this book

To build the projects you will find in this book, any experience with programming and/or electronics is of course appreciated, but you will be able to follow even if you have little experience in the field, as we will start from the absolute basics of the ESP8266.

Who this book is for

This book is for people who want to build their own Internet of Things projects, using the ESP8266 as the platform to easily build those projects.

It is also for people already building IoT projects, for example with Arduino, and those who want to discover another platform to build IoT projects.

Sections

In this book, you will find several headings that appear frequently (Getting ready, How to do it, How it works, There's more, and See also).

To give clear instructions on how to complete a recipe, we use these sections as follows:

Getting ready

This section tells you what to expect in the recipe, and describes how to set up any software or any preliminary settings required for the recipe.

How to do it...

This section contains the steps required to follow the recipe.

How it works...

This section usually consists of a detailed explanation of what happened in the previous section.

There's more...

This section consists of additional information about the recipe in order to make the reader more knowledgeable about the recipe.

See also

This section provides helpful links to other useful information for the recipe.

Conventions

In this book, you will find a number of text styles that distinguish between different kinds of information. Here are some examples of these styles and an explanation of their meaning.

Code words in text, database table names, folder names, filenames, file extensions, pathnames, dummy URLs, user input, and Twitter handles are shown as follows: "We will configure pin 5 as an input and then read it using the `digitalRead()` function and display the state of the input signal on the serial monitor."

A block of code is set as follows:

```
// LED pin
int inputPin = 5;
int val = 0;

void setup() {
  Serial.begin(9600);
  pinMode(inputPin, INPUT);
}

void loop() {

  // read pin
  val = digitalRead(inputPin);

  // display state of input pin
  Serial.println(val);
  delay(1000);
}
```

Any command-line input or output is written as follows:

```
esptool.py --port /dev/ttyUSB0 --baud 460800 write_flash --flash_
size=detect 0 esp8266-2016-05-03-v1.8.bin
```

New terms and **important words** are shown in bold. Words that you see on the screen, for example, in menus or dialog boxes, appear in the text like this: "The .bin files for the MicroPython port for ESP8266 are under the **Firmware for ESP8266 boards** sub heading."

 Warnings or important notes appear in a box like this.

 Tips and tricks appear like this.

Reader feedback

Feedback from our readers is always welcome. Let us know what you think about this book— what you liked or disliked. Reader feedback is important for us as it helps us develop titles that you will really get the most out of.

To send us general feedback, simply e-mail feedback@packtpub.com, and mention the book's title in the subject of your message.

If there is a topic that you have expertise in and you are interested in either writing or contributing to a book, see our author guide at www.packtpub.com/authors.

Customer support

Now that you are the proud owner of a Packt book, we have a number of things to help you to get the most from your purchase.

Downloading the example code

You can download the example code files for this book from your account at http://www.packtpub.com. If you purchased this book elsewhere, you can visit http://www.packtpub.com/support and register to have the files e-mailed directly to you.

You can download the code files by following these steps:

1. Log in or register to our website using your e-mail address and password.
2. Hover the mouse pointer on the **SUPPORT** tab at the top.
3. Click on **Code Downloads & Errata**.

4. Enter the name of the book in the **Search** box.

5. Select the book for which you're looking to download the code files.

6. Choose from the drop-down menu where you purchased this book from.

7. Click on **Code Download**.

Once the file is downloaded, please make sure that you unzip or extract the folder using the latest version of:

▶ WinRAR / 7-Zip for Windows

▶ Zipeg / iZip / UnRarX for Mac

▶ 7-Zip / PeaZip for Linux

The code bundle for the book is also hosted on GitHub at `https://github.com/PacktPublishing/ESP8266-Internet-of-Things-Cookbook`. We also have other code bundles from our rich catalog of books and videos available at `https://github.com/PacktPublishing/`. Check them out!

Downloading the color images of this book

We also provide you with a PDF file that has color images of the screenshots/diagrams used in this book. The color images will help you better understand the changes in the output. You can download this file from `http://www.packtpub.com/sites/default/files/downloads/ESP8266InternetofThingsCookbook_ColorImages.pdf`.

Errata

Although we have taken every care to ensure the accuracy of our content, mistakes do happen. If you find a mistake in one of our books—maybe a mistake in the text or the code—we would be grateful if you could report this to us. By doing so, you can save other readers from frustration and help us improve subsequent versions of this book. If you find any errata, please report them by visiting `http://www.packtpub.com/submit-errata`, selecting your book, clicking on the **Errata Submission Form** link, and entering the details of your errata. Once your errata are verified, your submission will be accepted and the errata will be uploaded to our website or added to any list of existing errata under the Errata section of that title.

To view the previously submitted errata, go to `https://www.packtpub.com/books/content/support` and enter the name of the book in the search field. The required information will appear under the **Errata** section.

Piracy

Piracy of copyrighted material on the Internet is an ongoing problem across all media. At Packt, we take the protection of our copyright and licenses very seriously. If you come across any illegal copies of our works in any form on the Internet, please provide us with the location address or website name immediately so that we can pursue a remedy.

Please contact us at `copyright@packtpub.com` with a link to the suspected pirated material.

We appreciate your help in protecting our authors and our ability to bring you valuable content.

Questions

If you have a problem with any aspect of this book, you can contact us at `questions@packtpub.com`, and we will do our best to address the problem.

1

Configuring the ESP8266

In this chapter, we will cover:

- ▸ Setting up the Arduino development environment for the ESP8266
- ▸ Choosing an ESP8266
- ▸ Required additional components
- ▸ Uploading your first sketch to the ESP8266
- ▸ Connecting the ESP8266 to your local Wi-Fi network
- ▸ Connecting the ESP8266 to a cloud server
- ▸ Troubleshooting basic ESP8266 issues

Introduction

This being the first chapter, we will be looking at how to get started with configuring the ESP8266 to connect to the Internet. This chapter will be the stepping stone to the much more fun and exciting projects that are in this book. Therefore, follow the instructions provided carefully.

In this chapter, you will learn how to set up the Arduino IDE and upload sketches to the ESP8266. You will also be guided on how to choose an ESP8266 module for your project and how to use the ESP8266 to connect to Wi-Fi networks and the Internet.

Setting up the Arduino development environment for the ESP8266

To start us off, we will look at how to set up an Arduino IDE development environment so that we can use it to program the ESP8266. This will involve installing the Arduino IDE and getting the board definitions for our ESP8266 module.

Getting ready

The first thing you should do is download the Arduino IDE if you do not already have it installed on your computer. You can do that from this link:

`https://www.arduino.cc/en/Main/Software`.

The web page will appear as shown. It features that latest version of the Arduino IDE. Select your operating system and download the latest version that is available when you access the link (it was 1.6.13 at the time of writing):

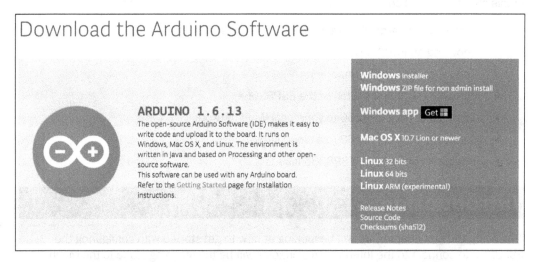

When the download is complete, install the Arduino IDE and run it on your computer.

Now that the installation is complete, it is time to get the ESP8266 definitions. Open the preference window in the Arduino IDE from **File | Preferences** or by pressing *CTRL + Comma*.

Copy this URL: `http://arduino.esp8266.com/stable/package_esp8266com_index.json`.

Paste it in the file labeled **Additional Board Manager URLs**, as shown in the following screenshot. If you are adding other URLs too, use a comma to separate them:

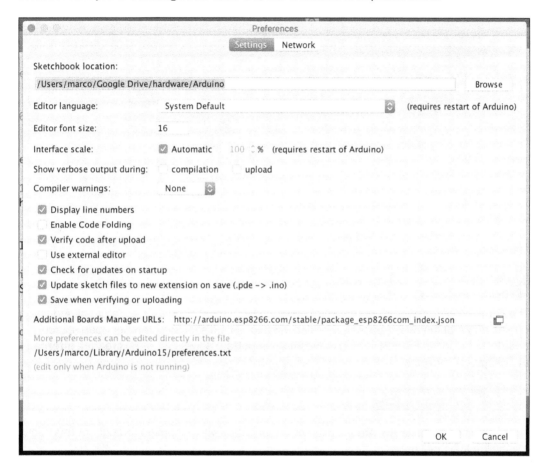

Open the board manager from the **Tools | Board menu** and install the ESP8266 platform. The board manager will download the board definition files from the link provided in the preferences window and install them. When the installation is complete, the ESP8266 board definitions should appear as shown in the screenshot. Now you can select your ESP8266 board from the **Tools | Board menu**:

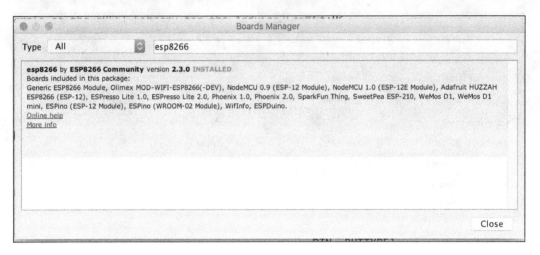

How it works...

The Arduino IDE is an open source development environment used for programming Arduino boards and Arduino-based boards. It is also used to upload sketches to other open source boards, such as the ESP8266. This makes it an important accessory when creating **Internet of Things (IoT)** projects.

See also

These are the basics for the Arduino framework and they will be applied throughout this book to develop IoT projects.

Choosing an ESP8266 board

The ESP8266 module is a self-contained **System On Chip** (**SOC**), which features an integrated TCP/IP protocol stack that allows you to add Wi-Fi capability to your projects. The module is usually mounted on circuit boards that break out the pins of the ESP8266 chip, making it easy for you to program the chip and to interface with input and output devices.

ESP8266 boards come in different forms, depending on the company that manufactures them. All the boards use Espressif's ESP8266 chip as the main controller, but they have different additional components and different pin configurations, giving each board unique additional features.

Therefore, before embarking on your IoT project, take some time to compare and contrast the different types of ESP8266 boards that are available. This way, you will be able to select the board that has features best suited for your project.

Available options

The simple ESP8266-01 module is the most basic ESP8266 board available on the market. It has eight pins, which include four **General Purpose Input/Output** (**GPIO**) pins, serial communication TX and RX pins, an enable pin and power pins, and VCC and GND. Since it only has four GPIO pins, you can only connect three inputs or outputs to it.

The 8-pin header on the ESP8266-01 module has a 2.0 mm spacing that is not compatible with breadboards. Therefore, you have to look for another way to connect the ESP8266-01 module to your setup when prototyping. You can use female to male jumper wires to do that:

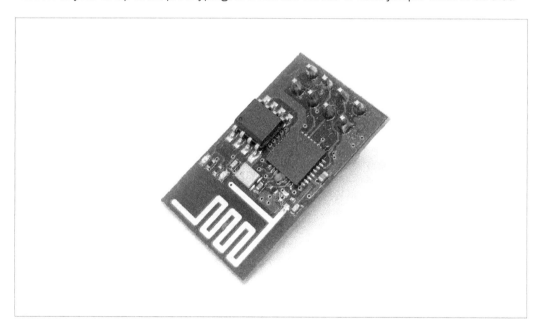

The ESP8266-07 is an improved version of the ESP8266-01 module. It has 16 pins, which consist of nine GPIO pins, serial communication TX and RX pins, a reset pin, an enable pin and power pins, and VCC and GND. One of the GPIO pins can be used as an analog input pin. The board also comes with a UFL. connector that you can use to plug an external antenna in case you need to boost Wi-Fi signal.

Since the ESP8266 has more **GPIO** pins, you can have more inputs and outputs in your project. Moreover, it supports both SPI and I2C interfaces, which can come in handy if you want to use sensors or actuators that communicate using any of those protocols. Programming the board requires the use of an external FTDI breakout board based on USB to serial converters, such as the FT232RL chip.

The pads/pinholes of the ESP8266-07 have a 2.0 mm spacing, which is not breadboard-friendly. To solve this, you have to acquire a plate holder that breaks out the ESP8266-07 pins to a breadboard-compatible pin configuration, with 2.54 mm spacing between the pins. This will make prototyping easier.

This board has to be powered from a 3.3V, which is the operating voltage for the ESP8266 chip:

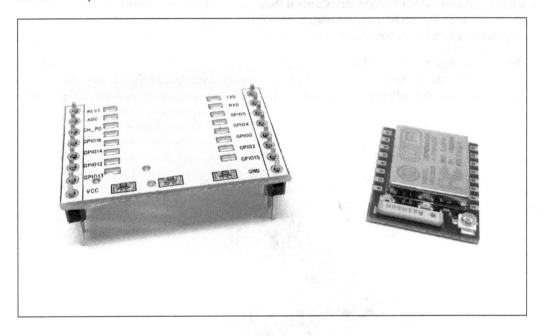

The Olimex ESP8266 module is a breadboard-compatible board that features the ESP8266 chip. As with the ESP8266-07 board, it has SPI, I2C, serial UART, and GPIO interface pins. In addition to that, it also comes with a **Secure Digital Input/Output** (**SDIO**) interface, which is ideal for communication with an SD card. This adds six extra pins to the configuration, bringing the total to 22 pins.

Since the board does not have an on-board USB to serial converter, you have to program it using an FTDI breakout board or a similar USB to serial board/cable. Moreover, it has to be powered from a 3.3V source, which is the recommended voltage for the ESP8266 chip:

The **Sparkfun ESP8266 Thing** is a development board for the ESP8266 Wi-Fi SOC. It has 20 pins that are breadboard-friendly, which makes prototyping easy. It features SPI, I2C, serial UART, and GPIO interface pins, enabling it to be interfaced with many input and output devices. There are eight GPIO pins, including the I2C interface pins.

The board has a 3.3V voltage regulator, which allows it to be powered from sources that provide more than 3.3V. It can be powered using a micro USB cable or Li-Po battery. The USB cable also charges the attached Li-Po battery, thanks to the Li-Po battery charging circuit on the board.

Programming has to be done via an external FTDI board:

The **Adafruit Huzzah ESP8266** is a fully standalone ESP8266 board. It has a built-in USB to serial interface that eliminates the need for using an external FTDI breakout board to program it. Moreover, it has an integrated battery charging circuit that charges any connected Li-Po battery when the USB cable is connected. There is also a 3.3V voltage regulator on the board that allows the board to be powered with more than 3.3V.

Though there are 28 breadboard friendly pins on the board, only 22 are useable. Ten of those pins are GPIO pins and they can also be used for SPI as well as I2C interfacing. One of the GPIO pins is an analog pin:

What to choose?

All the ESP8266 boards will add Wi-Fi connectivity to your project. However, some of them lack important features and are difficult to work with. So the best option would be to use the module that has the most features and is easy to work with. The Adafruit ESP8266 fits the bill.

The Adafruit ESP8266 is completely standalone and easy to power, program, and configure due to its on-board features. Moreover, it offers many input/output pins that will enable you to add more features to your projects. It is affordable and small enough to fit in projects with limited space.

There's more...

Wi-Fi isn't the only technology that we can use to connect our projects to the Internet. There are other options such as Ethernet and 3G/LTE. There are shields and breakout boards that can be used to add these features to open source projects. You can explore these other options and see which works for you.

See also

Now that we have chosen the board to use in our project, you can proceed to the next step, which is understanding all the components we will use with Adafruit ESP8266 in this book.

Required additional components

To demonstrate how the ESP8266 works, we will use some additional components. These components will help us learn how to read sensor inputs and control actuators using GPIO pins. Through this you can post sensor data to the Internet and control actuators from Internet resources, such as websites.

Required components

The components we will use include:

- ▶ Sensors:
    - ❏ DHT11
    - ❏ Photocell
    - ❏ Soil humidity

- ▸ Actuators:
    - ❏ Relay
    - ❏ Power switch tail kit
    - ❏ Water pump
- ▸ Breadboard
- ▸ Jumper wires
- ▸ Micro USB cable

Sensors

Let's discuss the three sensors that we will be using.

DHT11

The DHT11 is a digital temperature and humidity sensor. It uses a thermistor and capacitive humidity sensor to monitor the humidity and temperature of the surrounding air, and produces a digital signal on the data pin. A digital pin on the ESP8266 can be used to read the data from the sensor data pin.

 The DHT11 sensor is not very precise, but it is perfect for experimenting, which we'll be doing in this book.

Photocell

A photocell is a light sensor that changes its resistance depending on the amount of incident light it is exposed to. They can be used in a voltage divider setup to detect the amount of light in the surroundings. In a setup where the photocell is used in the VCC side of the voltage divider, the output of the voltage divider goes high when the light is bright and low when the light is dim. The output of the voltage divider is connected to an analog input pin and the voltage readings can be read:

Soil humidity sensor

The soil humidity sensor is used for measuring the amount of moisture in soil and other similar materials. It has two large exposed pads that act as a variable resistor. If there is more moisture in the soil, the resistance between the pads drops, leading to a higher output signal. The output signal is connected to an analog pin from where its value is read.

 This sensor is mainly used for demonstration purposes, but it is perfect for the projects we'll do in this book.

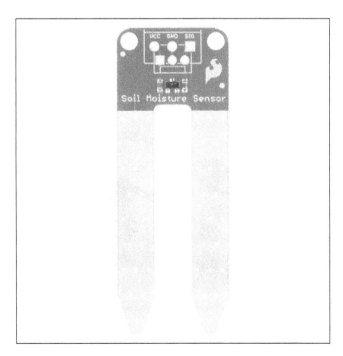

Actuators

Let's discuss the actuators.

Relays

A relay is a switch that is operated electrically. It uses electromagnetism to switch large loads using small voltages. It consists of three parts: a coil, spring, and contacts. When the coil is energized by a high signal from a digital pin on the ESP8266, it attracts the contacts, forcing them closed. This completes the circuit and turns on the connected load. When the signal on the digital pin goes low, the coil is no longer energized and the spring pulls the contacts apart. This opens the circuit and turns off the connected load:

Power switch tail kit

A power switch tail kit is a device that is used to control standard wall outlet devices with microcontrollers. It is already packaged to prevent you from having to mess around with high voltage wiring. Using it, you can control appliances in your home using the ESP8266:

Water pump

A water pump is used to increase the pressure of fluids in a pipe. It uses a DC motor to rotate a fan and create a vacuum that sucks up the fluid. The sucked fluid is then forced to move by the fan, creating a vacuum again that sucks up the fluid behind it. This in effect moves the fluid from one place to another:

Breadboard

A breadboard is used to temporarily connect components without soldering. This makes it an ideal prototyping accessory that comes in handy when building circuits:

Jumper wires

Jumper wires are flexible wires that are used to connect different parts of a circuit on a breadboard:

Micro USB cable

A micro USB cable will be used to connect the Adafruit ESP8266 board to the computer:

See also

Having understood the components that you are going to use with Adafruit ESP8266, you can now proceed to the next step and learn how to upload sketches to your ESP8266.

Uploading your first sketch to the ESP8266

In this recipe, we are going to look at how to upload sketches to the ESP8266. This will enable you to understand some basics of the Arduino programming language and get you to upload your first sketch to your ESP8266 board.

Getting ready

Ensure that the board definitions of the ESP8266 board are installed in your Arduino IDE as explained earlier, and then connect your ESP8266 board to the computer. Now you can proceed to do the rest of the connections.

In this project, you will require a few extra components. They are:

- LED (https://www.sparkfun.com/products/528)
- 220 Ω resistor (https://www.sparkfun.com/products/10969)
- Breadboard
- Jumper wires

Start by mounting the LED onto the breadboard. Connect one end of the 220 Ω resistor to the positive leg of the LED (the positive leg of an LED is usually the taller one of the two legs). Connect the other end of the resistor on another rail of the breadboard and connect one end of the jumper wire to that rail and the other end of the jumper wire to pin 5 of the ESP8266 board. Take another jumper wire and connect one if its ends to the negative leg of the LED and connect the other end to the GND pin of the ESP8266.

How to do it...

We will configure the board to output a low signal on pin 5 for a duration of one second and then output a high signal on pin 5 for a duration of two seconds. The process will repeat forever:

```
// LED pin
int ledPin = 5;

void setup() {
  pinMode(ledPin, OUTPUT);
}

void loop() {

  // OFF
  digitalWrite(ledPin, LOW);
  delay(1000);
```

```
    // ON
    digitalWrite(ledPin, HIGH);
    delay(2000);
}
```

Refer the following steps:

1. Copy the code and paste it in your Arduino IDE.

2. Ensure your ESP8266 board is connected to your computer, and then proceed to select the board in **Tools|Board menu**. If the ESP8266 board definitions were properly installed, you should see a list of all ESP8266 boards in the menu:

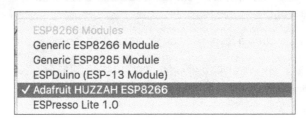

3. Choose the board type you are using. In this case it is the **Adafruit HUZZAH ESP8266**.

4. Select the serial port where the ESP8266 board is connected from the **Tools|Port menu** and then proceed to upload the code to your board:

How it works...

This is a simple LED blinking program that we used to demonstrate how to upload code to an Adafruit ESP8266 board. The program blinks an LED connected to pin 5 of the Adafruit ESP8266 board every three seconds. This is done by turning off the LED for one second and turning on the LED for two seconds, continuously.

There's more...

The frequency at which the LED blinks can be increased or reduced by adjusting the delays in the program. For instance, the LED can be made to blink faster through reducing the second delay from two seconds to one second, by changing this statement `delay(2000);` to `delay(1000);`.

See also

Having successfully uploaded your first sketch to your ESP8266 board, you can go to the next recipe and learn how to connect your ESP8266 board to a local Wi-Fi network.

Connecting the ESP8266 to your local Wi-Fi network

In this recipe, we will show you how to connect your ESP8266 board to your local Wi-Fi network. Before you can send or receive data to and from the Internet, your ESP8266 board has to be connected to a Wi-Fi network that has Internet connectivity. This is done in the setup part of your program since it is supposed to be done once, when the ESP8266 is powered on.

Getting ready

Ensure your ESP8266 board is connected to your computer via the USB cable. No other components will be used in this exercise, since there are no inputs or outputs that are required.

How to do it...

We will include the `ESP8266` library in the program and set the Wi-Fi network `ssid` and `password` so that the `ESP8266` connects to the network when it gets powered on. We will print out a confirmation and the IP address of the `ESP8266` when the connection is successful.

This is the sketch for this exercise:

```
// Libraries
#include <ESP8266WiFi.h>

// WiFi network
const char* ssid     = "your-ssid";
const char* password = "your-password";

void setup() {

  // Start serial
```

```
    Serial.begin(115200);
    delay(10);

    // Connecting to a WiFi network
    Serial.println();
    Serial.println();
    Serial.print("Connecting to ");
    Serial.println(ssid);

    WiFi.begin(ssid, password);

    while (WiFi.status() != WL_CONNECTED) {
      delay(500);
      Serial.print(".");
    }

    Serial.println("");
    Serial.println("WiFi connected");
    Serial.println("IP address: ");
    Serial.println(WiFi.localIP());
  }

  void loop() {

  }
```

Refer the following steps:

1. Copy this sketch and paste it in your Arduino IDE.

2. Change the SSID in the code from `your-ssid` to the name of your Wi-Fi network, and the password from `your-password` to the password of your Wi-Fi network.

3. Make sure you have selected the correct board from the **Tools | Board menu**, which in this case is **Adafruit HUZZAH ESP8266**, and the correct serial port from the **Tools | Port menu**.

4. Upload the code to your ESP8266 board.

5. Open the serial monitor (and make sure that the serial speed is set at 115200) so that you can check to see whether the ESP8266 board has connected to the Wi-Fi network.

The results are as shown in the following screenshot. The ESP8266 prints its IP address once the connection has been successfully established:

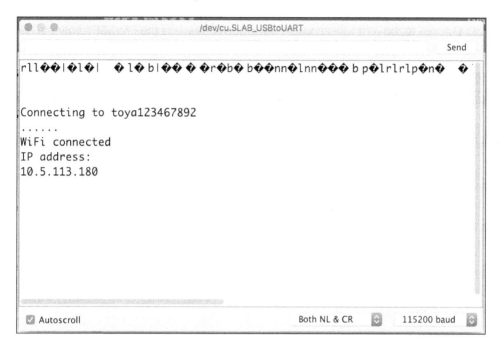

How it works...

The program sets up the Wi-Fi SSID and password using the `WiFi.begin()` function of the `ESP8266Wifi` library, and then executes the `WiFi.status()` function to try to connect to the Wi-Fi network. The program checks whether the `WiFi.status()` function returns a `true` value to indicate that the ESP8266 board has successfully connected to the Wi-Fi network. If the connection was not successful, the sketch tries to connect again using the `WiFi.status()` function and repeats that until the `WiFi.status()` function returns a `true` value to indicate that the ESP8266 has successfully connected to the Wi-Fi network.

When the ESP8266 connects to the Wi-Fi network, the sketch displays a message, `WiFi connected`, and the IP address of the ESP8266 on the serial monitor.

There's more...

You can enter the wrong password and SSID and see what output you get in the serial monitor.

Now that you have successfully connected your ESP8266 board to your local Wi-Fi network, you can proceed to the next recipe, where we will be connecting the ESP8266 board to the Internet.

Connecting the ESP8266 to a cloud server

In this recipe, we will connect the ESP8266 to the Internet and send data to a cloud server. The cloud server we will be sending data to is dweet.io. The data we send to dweet.io will be used later in this book, so ensure that you execute this section successfully.

Getting ready

As in the previous recipe, we won't need any extra components here. All we need to do is ensure that the ESP8266 is connected to the computer.

How to do it...

To accomplish this, follow these steps:

1. We will connect the ESP8266 to a local Wi-Fi network that has an active Internet connection.

2. Once the connection is successful, we will send a GET request to the cloud server and then display the reply that the server sends back to the ESP8266 board:

```
// Libraries
#include <ESP8266WiFi.h>
```

3. Enter the SSID and password:

```
// SSID
const char* ssid     = "your-ssid";
const char* password = "your-password";
```

4. Store the hostname of the cloud server:

```
// Host
const char* host = "dweet.io";
```

5. Configure the SSID and password of the Wi-Fi network and connect the ESP8266 to the Wi-Fi network:

```
void setup() {

  // Serial
  Serial.begin(115200);
```

```
delay(10);

// We start by connecting to a WiFi network
Serial.println();
Serial.println();
Serial.print("Connecting to ");
Serial.println(ssid);

WiFi.begin(ssid, password);

while (WiFi.status() != WL_CONNECTED) {
  delay(500);
  Serial.print(".");
}

Serial.println("");
Serial.println("WiFi connected");
Serial.println("IP address: ");
Serial.println(WiFi.localIP());
}
```

6. Delay for five seconds and then print the name of the host we are connecting to on the serial monitor:

```
void loop() {

  delay(5000);

  Serial.print("connecting to ");
  Serial.println(host);
```

7. Connect to the host server:

```
// Use WiFiClient class to create TCP connections
WiFiClient client;
const int httpPort = 80;
if (!client.connect(host, httpPort)) {
  Serial.println("connection failed");
  return;
}
```

8. Formulate the URI for the GET request we will send to the host server:

```
// We now create a URI for the request
String url = "/dweet/for/my-thing-name?value=test";
```

9. Send the GET request to the server and check whether the request has been received or if it has timed out:

```
// Send request
Serial.print("Requesting URL: ");
Serial.println(url);

client.print(String("GET ") + url + " HTTP/1.1\r\n" +
             "Host: " + host + "\r\n" +
             "Connection: close\r\n\r\n");
unsigned long timeout = millis();
while (client.available() == 0) {
  if (millis() - timeout > 5000) {
    Serial.println(">>> Client Timeout !");
    client.stop();
    return;
  }
}
```

10. Read incoming data from the host server line by line and display the data on the serial monitor.

11. Close the connection after all the data has been received from the server:

```
// Read all the lines from the answer
while(client.available()){
  String line = client.readStringUntil('\r');
  Serial.print(line);
}

// Close connecting
Serial.println();
Serial.println("closing connection");
}
```

12. Copy the sketch to your Arduino IDE and change the SSID in the code from your-ssid to the name of your Wi-Fi network and the password from your-password to the password of your Wi-Fi network.

13. Upload the sketch to your ESP8266 board.

14. Open the serial monitor so that you can view the incoming data.

The serial monitor should display data, as shown in the following screenshot:

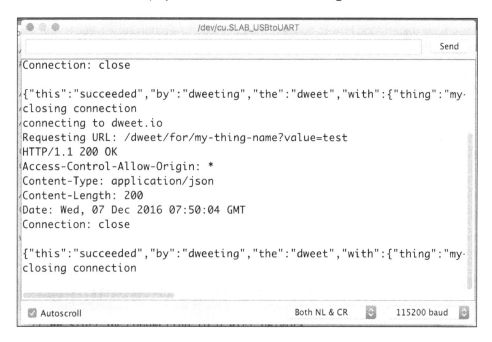

As you can see from the serial monitor, when you send the GET request to dweet.io you receive a reply from the server. The reply is enclosed in curly brackets { }.

How it works...

The program connects to the Wi-Fi network using the provided password and SSID. It then proceeds to connect to the provided cloud/host server using the `client.connect()` function, and sends the provided URI to the host server using the `client.print()` function.

Once the data has been successfully sent, the sketch waits for a reply from the server. It does this with the `client.available()` function, which checks whether there is incoming data from the server. If there is data available, the sketch reads it and displays it on the serial monitor. The process is repeated until the ESP8266 is turned off.

There's more...

Since you have understood how to connect the ESP8266 to a cloud server, see if you can change the sketch so that it connects to the www.google.com host server and searches for the word Arduino. The results from the server should be displayed on the serial monitor.

 Hint: You can check out the following link to see the format for Google GET requests:

https://www.google.com/support/enterprise/static/gsa/docs/admin/72/gsa_doc_set/xml_reference/request_format.html.

See also

You can now continue to the next recipe and look at some of the common issues that you may face when using your ESP8266 and how you can solve them.

Troubleshooting basic ESP8266 issues

There are several issues you may face when using ESP8266 boards. Here are some of the common ones and ways to solve them.

The Board is not visible from the Arduino IDE

One of the causes of this problem is a faulty USB cable. You can troubleshoot this by using another USB cable to see whether it works. If it works, then your USB cable is faulty and should be replaced. If the problem still persists, even when you are using another USB cable, the issue may be with the board definitions.

This issue can also occur when the board definitions have not been installed properly, or if they have not been installed at all. So make sure that the board definitions have been installed properly, or try and reinstall them again.

The board cannot be configured from the Arduino IDE

A faulty USB cable can cause this kind of issue. To troubleshoot, use another USB cable and see whether the issue persists. If it persists, keep your USB cable and try another solution. However, if the issue gets solved, it means that your USB cable is indeed faulty and should be replaced.

If the USB cable is not faulty but the issue continues, press the reset button on the board. That should deal with the problem.

The board does not connect to your local Wi-Fi network

If your board does not connect to your local Wi-Fi network, check your Wi-Fi network's security protocol and confirm it is WPA. If it is not, change it to WPA. Moreover, confirm the password you have written in the sketch is the correct one. Also avoid using Wi-Fi passwords with strange characters. It could also be that the serial port is not set to the right speed, therefore you might not see the Wi-Fi connected message.

2
Your First ESP8266 Projects

In this chapter, we will cover:

- ▶ Functionalities of the ESP8266
- ▶ Reading digital signals
- ▶ Reading analog signals
- ▶ Controlling an LED
- ▶ Dimming an LED
- ▶ Controlling a servo motor
- ▶ Measuring data from a digital sensor
- ▶ Controlling an OLED screen
- ▶ Troubleshooting usual issues with ESP8266 basics

Introduction

In this second chapter of the book, we will learn how to use the **General Purpose Input/ Output** (**GPIO**) pins to read input signals and produce output signals. With that knowledge,we will create simple projects to control outputs and read sensor inputs. This will give way to the development of more complex projects in later chapters.

Functionalities of the ESP8266

The first thing we will tackle is the functionality of the ESP8266. We will look at the features and functions of the ESP8266, and learn about all the available GPIO pins and how we can use them to read and write digital signals.

Features

The ESP8266 chip provides a self-contained, standalone Wi-Fi networking solution that allows you to host applications and to provide a Wi-Fi network that you can use to transfer data and instructions in your projects.

In normal use, the ESP8266 boots up directly from an external flash memory. It also has a cache that improves the performance of the system and reduces the amount of memory needed during operation.

The ESP8266 can also be used as a Wi-Fi adapter. In this case, the chip provides wireless Internet access to a microcontroller-based project through either the CPU AHB bridge or UART interface.

The on-board processing and storage capabilities of the ESP8266 enable easy integration with sensors and other devices via the GPIO pins. Moreover, the ESP8266 requires little to no external circuitry because of its on-chip integration, which includes power management converters and antenna switch balun. This makes it small and ideal for compact projects.

In addition to all those features, the ESP8266 has some sophisticated system-level features that include adaptive radio biasing for low-power operation, energy-efficient VoIP due to rapid sleep/wake switching, advanced signal processing and radio co-existence, and spur cancellation features that prevent cellular and Bluetooth interference.

The main features of the ESP8266 can be listed as:

- 802.11 b/g/n protocol
- Integrated TCP/IP protocol stack
- Soft-AP, Wi-Fi direct (P2P)
- Integrated power amplifier an management units, PLL, TR switch, regulators, LNA, switch, matching network, and balun
- Integrated 32-bit low-power CPU used as the application processor
- Supports antenna diversity
- Supports UART, SPI, and SDIO 2.0 interfaces
- 2 ms wake up and transmit interval

- ▶ Power down leakage current of less than 10 µA
- ▶ Standby power of less than 1.0 mW
- ▶ Output power of +19.5 dBm when using 802.11 mode
- ▶ STBC, 2x1 MIMO, 1x1 MIMO
- ▶ A-MSDU and A-MPDU aggregation

Pin configuration

To understand the ESP8266 pin configuration, we are going to look at the ESP-12 module. This is because it is the same module that is used on the Adafruit feather HUZZAH ESP8266 Wi-Fi board that we are using for our exercises, as explained in *Chapter 1, Configuring the ESP8266*. The following diagram shows the ESP-12 pin configuration:

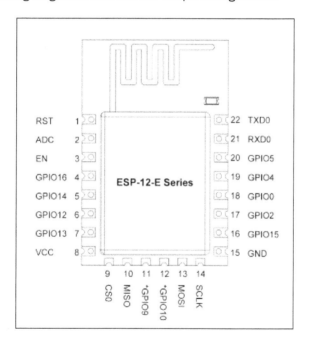

The ESP-12 module pins are connected to a more breadboard-friendly pin configuration on the Adafruit ESP8266 board. The pin configuration of the Adafruit board is shown in the following diagram:

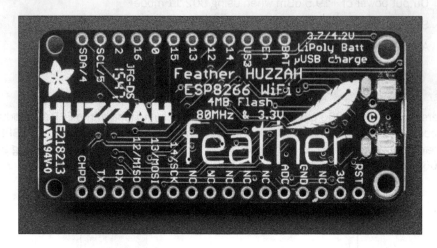

The pins on the Adafruit ESP8266 module can be grouped into several categories.

Power pins

The power pins are:

- **BAT**: Positive voltage from the Li-Po battery that is connected to the JST jack
- **GND**: Common ground for the power and logic
- **USB**: Positive voltage from the micro USB port on the board
- **3V**: Output voltage from the on-board 3.3V regulator, which supplies a maximum of 500 mA
- **EN**: 3.3V voltage regulator enable pin; to disable the 3.3V voltage regulator, connect this pin to GND

Serial pins

The serial pins are labelled RX (receive) and TX (transmit). Note that those pins can also be used for I2S communications. They are used for serial communication and boot loading. The RX pin is the input pin and is 5V compliant, since it has a level shifter. The TX pin is the output from the module and has a 3.3V logic. The two pins are connected to the CP2104 USB to serial converter and should be left disconnected unless you really want to use them.

I2C and SPI pins

The ESP8266 module comes with both I2C and SPI interfaces, which enable you to communicate with sensors, outputs, and devices that use these interfaces. Though this is done through bitbanging, it works just as well as it does on the Arduino.

Hypothetically, any GPIO pin on the ESP8266 can be used for I2C and SPI, but the current library for the Adafruit ESP8266 uses the following pins:

- For I2C:
    - I2C SCL – GPIO 5
    - I2C SDA – GPIO 4

- For SPI:
    - SPI SCK – GPIO 14
    - SPI MISO – GPIO 2
    - SPI MOSI – GPIO 13

GPIO pins

The **Adafruit HUZZAH ESP8266** has nine GPIO pins that you can use. They are: 0, 2, 4, 5, 12, 13, 14, 15, and 16.

They all have a 3.3V logic level and are not 5V compatible. The maximum current that can be drawn from one pin is 12 mA. The pins are general purpose and can be used for input as well as output. Additional functions of the pins are explained following:

- GPIO 0 does not have an internal pullup resistor and is connected to a red LED which it controls. In addition to this, the pin is used to set the ESP8266 into bootloader mode. This is achieved by pulling the pin low during power-up.

- GPIO 2 has a pullup resistor and is connected to the blue LED, that is close to the Wi-Fi antenna. It is used to detect boot mode and as an output to control the blue LED.

- GPIO 15, the same as pin 2, is used to detect boot mode. There is a pulldown resistor connected to it. Always ensure this pin isn't pulled high during power-up. It can only be used as an output.

- GPIO 16 is used to wake up the module from deep sleep mode. It is advisable to connect it to the reset pin.

[GPIO pins 12, 13, and 14 are also used for SPI interfacing.]

Analog pins

There is only one analog pin on the ESP8266 board. It is labelled ADC. The pin can read a maximum voltage of approximately 1.0V. If the voltage you want to read is higher than 1.0V, you have to use a voltage divider to convert it to a 0-1.0V range.

Control pins

There are two control pins available on the ESP8266 board. They are:

  ▸ RST – Reset pin for ESP8266. It is always pulled high and works with 3.3V logic only. When it's connected to ground it resets the ESP8266.

  ▸ EN (CH_PD) – Enable pin for the ESP8266. It is always pulled high and works with 3.3V logic. When pulled to ground, it resets the ESP8266.

The pins labelled NC are not connected to anything and are only used as placeholders. They make the Adafruit feather HUZZAH ESP8266 board compatible with other feather boards manufactured by Adafruit.

How it works...

The ESP8266 module reads digital signals from inputs and writes digital signals to outputs via the GPIO pins. Since all the GPIO pins have a logic voltage of 3.3V, it is important to ensure that any input signal to the pins is between 0V and 3.3V.

The GPIO pins usually read or write two kinds of digital signal: 0s and 1s. 0s represents voltages between -0.3V and 0.825V, while 1s represents voltages between 2.475V and 3.3V. Therefore, whenever you read a digital pin that is connected to a voltage above 2.475V, you get a 1 (high signal), and when you read a pin that is connected to a voltage less than 0.875V, you get a 0 (low signal), and the same applies to outputs.

See also

Now that you have understood the different features and the pin configuration of the ESP8266 module, you can go ahead and put that knowledge into action. Therefore, in the next recipe we will look at how to read digital inputs.

Reading digital signals

One of the uses of the ESP8266's GPIO pins is to read digital signals. This allows you to control your project using input devices and also to monitor sensor data. In this recipe, we will look at how to read digital signals using GPIO pins.

Getting ready

Connect your ESP8266 board to your computer via a USB cable and set up your Arduino IDE (refer back to *Chapter 1, Configuring the ESP8266*). Once that is done, you can proceed to make the other connections.

In this recipe, we will need a few extra components. They include:

▸ Breadboard
▸ Jumper wires

Mount the ESP8266 board onto the breadboard and then connect a jumper wire from pin **5** to the GND pin. The connection should be as shown in the following figure:

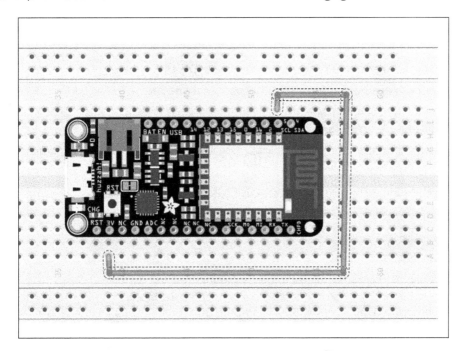

How to do it...

We will configure pin **5** as an input, then read it using the `digitalRead()` function and display the state of the input signal on the serial monitor. This will be repeated every 1 second:

```
// LED pin
int inputPin = 5;
int val = 0;
```

```
void setup() {
  Serial.begin(9600);
  pinMode(inputPin, INPUT);
}

void loop() {

  // read pin
  val = digitalRead(inputPin);

  // display state of input pin
  Serial.println(val);
  delay(1000);
}
```

Refer the following steps:

1. Copy the sketch and paste it in your Arduino IDE.
2. Check to ensure that the ESP8266 board is connected.
3. Select the board that you are using in **Tools | Board menu** (in this case it is the **Adafruit HUZZAH ESP8266**).
4. Select the serial port your board is connected to from the **Tools | Port menu** and then upload the code.

When you open your serial monitor, a zero will be displayed after one second and another after two seconds, and so on and so forth.

How it works...

The program initializes two variables at the beginning. The first variable (inputPin) is the GPIO pin number and the second variable (val) will hold the state of the GPIO pin. The serial communication baud rate is set at 9600 and GPIO 5 is configured as an input pin in the setup section of the sketch. In the loop section, the program reads the digital input on pin 5 and stores the value in the val variable, then displays the value stored in the val variable on the serial monitor. There is a delay of 1 second after that.

There's more...

If you have successfully completed that exercise, you can spice things up by connecting GPIO 5 to the 3.3V pin instead of the GND pin, then checking the serial monitor. You should see 1s listed on the serial monitor instead of 0s, since the input signal on pin 5 is high.

See also

Having successfully learned how to read digital signals, you can now proceed to the next recipe, where we will advance to *Reading analog signals*.

Reading analog signals

The ESP8266 has one analog pin that we can use to read analog signals. In this recipe, we will be looking at how to write a sketch that reads analog signals. This will enable us to read input from analog sensors.

Getting ready

Connect your ESP8266 board to your computer via a USB cable and set up your Arduino IDE (refer back to *Chapter 1, Configuring the ESP8266*). Once that is done, you can proceed to make the other connections.

In this recipe, we will need a breadboard and jumper wires in addition to the ESP8266 board. Mount the ESP8266 board onto the breadboard and then connect a jumper wire from the analog **ADC** pin to the **GND** pin. The connection should be as shown in the following diagram:

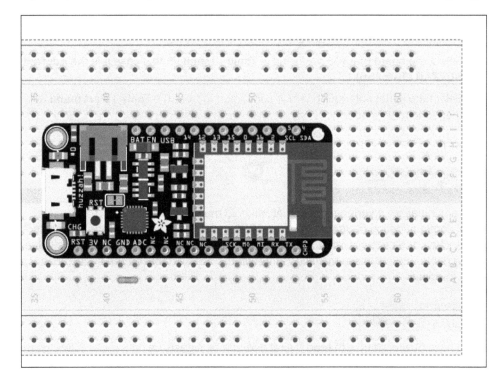

How to do it...

We will use the `analogRead()` function to read the analog signal on the **ADC** pin and display the analog signal value on the serial monitor. This will be repeated every 1 second:

```
// LED pin
int val = 0;

void setup() {
  Serial.begin(9600);
}

void loop() {

  // read pin
  val = analogRead(A0);

  // display state of input pin
  Serial.println(val);
  delay(1000);
}
```

1. Copy the sketch and paste it in your Arduino IDE.

2. Check to ensure that the ESP8266 board is connected.

3. Select the board that you are using in **Tools | menu** (in this case it is the **Adafruit HUZZAH ESP8266**).

4. Select the serial port your board is connected to from the **Tools | Port menu** and then upload the code.

When you open your serial monitor, you will notice that a zero is displayed every second.

How it works...

The program initializes a variable (`val`) that will hold the value of the read analog signal. The serial communication baud rate is set at 9600 in the setup section of the sketch. In the loop section, the program reads the analog input on the **ADC** pin and stores the value in the `val` variable, then displays the value stored in the `val` variable on the serial monitor at 1 second intervals.

See also

Since you have mastered how to read digital and analog signals using the ESP8266, go the next chapter and learn how to control outputs with the ESP8266.

Controlling an LED

This recipe is going to look at how to control an LED using an ESP8266 board. This will basically involve changing the state of an LED either ON or OFF, using the ESP8266 board's GPIO pins. The exercise will enable you to understand how to use the digital output function on the ESP8266.

Getting ready

Connect your ESP8266 board to your computer via a USB cable and set up your Arduino IDE (refer back to *Chapter 1, Configuring the ESP8266*). Once that is done, you can proceed to make the other connections.

In this recipe, we will need the following components:

- ESP8266 board
- USB cable
- LED (https://www.sparkfun.com/products/528)
- 220 Ω resistor (https://www.sparkfun.com/products/10969)
- Breadboard
- Jumper wires

Start by mounting the LED onto the breadboard. Connect one end of the 220 Ω resistor to the positive leg of the LED (the positive leg of an LED is usually the taller one of the two legs). Connect the other end of the resistor to another rail of the breadboard and connect one end of the jumper wire to that rail and the other end of the jumper wire to pin 5 of the ESP8266 board. Take another jumper wire and connect one of its ends to the negative leg of the LED and connect the other end to the **GND** pin of the ESP8266. The connection is as shown in the following diagram:

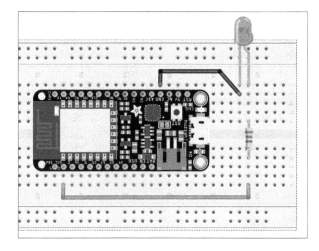

We will use the `digitalWrite()` function to output a HIGH signal on pin 5 for a duration of 1 second and then output a low signal on pin 5 for a duration of one second. This will be repeated over and over again to blink the LED:

```
// LED pin
int ledPin = 5;

void setup() {
  pinMode(ledPin, OUTPUT);
}

void loop() {

  // ON
  digitalWrite(ledPin, HIGH);
  delay(1000);

  // OFF
  digitalWrite(ledPin, LOW);
  delay(1000);
}
```

1. Copy the sketch and paste it in your Arduino IDE.
2. Check to ensure that the ESP8266 board is connected. Select the board that you are using in the **Tools | Board menu** (in this case it is the **Adafruit HUZZAH ESP8266**).
3. Select the serial port your board is connected to from the **Tools | Port menu** and then upload the code.

How it works...

The program uses the `digitalWrite()` function to change the state of the output pin, which in turn changes the state of the LED. Pin 5 is configured as an output in the setup section of the program. In the loop section, pin 5's state is set to HIGH (digital state 1) for one second. This turns the LED on for one second, after which pin 5's state is set to LOW (digital state 0) for one second, which turns the LED off for one second.

The `digitalWrite()` function is used to control the state of an output pin. The syntax is `digitalWrite(pin number, state)`. The pin number is the number assigned to the GPIO pin that is being controlled, and the state is the digital output signal (0 or 1). Instead of writing 1 and 0 as the state, we use `HIGH` and `LOW` respectively.

There's more...

Try to replace the `HIGH` and `LOW` states with 1 and 0 in the sketch and test it. You can also vary the delays in the sketch to change how the LED blinks.

See also

There is more you can do with LEDs, apart from just turning them on and off. In the next tutorial, we will be looking at how to dim LEDs using the ESP8266.

Dimming an LED

In this section, we are going to look at how to dim an LED using the analog output function. This will involve controlling the brightness of an LED that is attached to one of the GPIO pins of the ESP8266.

Getting ready

You will need an ESP8266 board and a USB cable to do this tutorial. There will also be some additional components required for this project. They include:

- LED (https://www.sparkfun.com/products/528)
- 220 Ω resistor (https://www.sparkfun.com/products/10969)
- Breadboard
- Jumper wires

Start by mounting the LED onto the breadboard. Connect one end of the 220 Ω resistor to the positive leg of the LED (the positive leg of an LED is usually the taller one of the two legs). Connect the other end of the resistor to another rail of the breadboard and connect one end of the jumper wire to that rail and the other end of the jumper wire to pin 4 of the ESP8266 board. Take another jumper wire and connect one of its ends to the negative leg of the LED and connect the other end to the **GND** pin of the ESP8266. The connection is as shown in the following diagram:

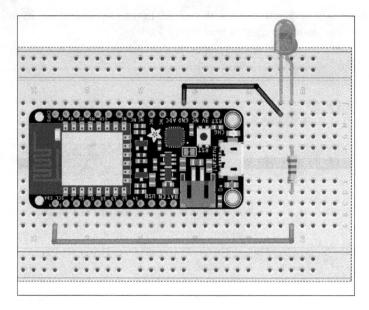

How to do it...

We will use the `analogWrite()` function to gradually reduce the duty cycle of the output signal to dim the LED. This will in turn reduce the brightness of the LED slowly until it completely turns off:

```
int ledPin = 4; // LED pin
int fadeValue = 1023; //duty cycle

void setup() {
  pinMode(ledPin, OUTPUT);
}

void loop() {
  analogWrite(ledPin, fadeValue); // glow LED
  if(fadeValue > 0) fadeValue --;  // decrease duty cycle by 1
  delay(5);
}
```

1. Copy the sketch and paste it in your Arduino IDE.

2. Check to ensure that the ESP8266 board is connected.

3. Select the board that you are using in the **Tools | Board menu** (in this recipe it is the **Adafruit HUZZAH ESP8266**).

4. Select the serial port your board is connected to from the **Tools | Port menu** and then upload the code.

How it works...

The program uses **Pulse Width Modulation** (**PWM**) to vary the brightness of the LED. This is implemented by the `analogWrite()` function. In the sketch, the function syntax is `analogWrite(pin, value)`. Pin refers to the ESP8266 board pin where the LED is connected to, which in our case is GPIO pin 4. Value refers to the duty cycle. It ranges between 0 and 1023. When the duty cycle is 0, the LED does not glow, and when it is 1023, the LED glows brightest.

In the sketch, GPIO pin 4 is configured as an output and the duty cycle is set at 1023. In the loop section of the sketch, the `analogWrite()` function glows the LED with the current duty cycle, then the duty cycle is reduced by 1 if it is not equal to 0 and then there is a delay of 5 milliseconds before the process is repeated again. This goes on until the duty cycle gets to 0.

When the sketch runs, the LED starts off at full brightness and then slowly dims until it goes off, over a time span of approximately 5 seconds.

There's more...

For further practice, edit the code so that instead of dimming the LED from full brightness to an off state, it starts from an off state and glows brighter until it reaches full brightness.

See also

Having understood how to control LEDs, you can now proceed to more complex output devices. The next recipe will take you through the basics of *Controlling a servo motor*.

Controlling a servo motor

In this section, we are going to look at how to use the functionalities of the ESP8266 to control a servo motor. We will be rotating the servo motor to a specific position in either direction at a set speed. This will demonstrate the different servo motor parameters we can control using an ESP8266 board.

Getting ready

Make sure you have all the components before proceeding. They include:

- ESP8266 board
- Mini USB cable
- Micro servo motor (`https://www.sparkfun.com/products/9065`)
- Breadboard
- Jumper wires

Start by mounting the ESP8266 board onto the breadboard. Use jumper cables to connect the power wires of the servo to the ESP8266 power pins. The power wires of the servo are coloured red for the positive terminal and black/brown for the negative terminal.

Connect the positive terminal to the USB pin of the ESP8266 board and the negative terminal to the GND pin of the ESP8266 board. We connect the positive terminal to the USB pin because it provides 5V, which is within the recommended servo motor power supply of 4.8V-6.0V.

Connect the signal wire of the servo motor to GPIO pin 2 on the ESP8266 board. The signal wire usually comes in different colors, depending on the servo motor. The most common colors are blue, orange, white, and yellow. You can also identify the signal wire through elimination, since it is the only other wire that is available apart from the two power supply wires.

Once all the connections are made, connect the ESP8266 board to the computer using the USB cable.

The setup is shown in the following diagram:

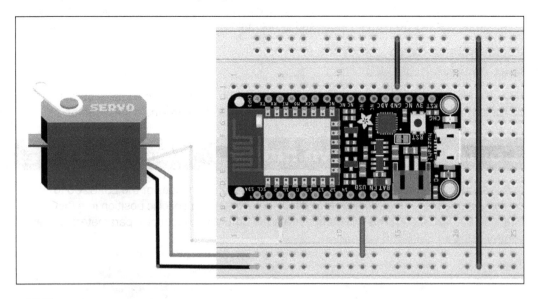

How to do it...

To control the servo motor, we will use PWM. The good news is that there is already a `servo motor` library for Arduino that handles all the hard PWM code and provides us with simple functions that we can use to control our servo motor.

With the `servo` library, we will use the `attach()` function to define the signal pin that the servo motor is connected to, and the `write()` function to instruct the servo to move to a specified position. To demonstrate how the two functions are used, we will move the servo motor position from 0 degrees to 180 degrees and then back to 0 degrees, and repeat this forever:

```
#include <Servo.h>

Servo myservo;  // create servo object to control a servo

void setup(){
  myservo.attach(2);  // attach the servo on GPIO 2
}

void loop(){
  int pos; // holds the position the servo should move to

  // goes from 0 degrees to 180 degrees
  // in steps of 1 degree
  for(pos = 0; pos <= 180; pos += 1){
    myservo.write(pos); // move servo to position in var pos
    delay(15);          // waits 15ms to reach the position
  }

  // goes from 180 degrees to 0 degrees
  // in steps of 1 degree
  for(pos = 180; pos>=0; pos-=1){
    myservo.write(pos); // move servo to position in var pos
    delay(15);          // waits 15ms to reach the position
  }
}
```

1. Copy the sketch and paste it in your Arduino IDE.
2. Check to ensure that the ESP8266 board is connected to the computer.
3. Select the board that you are using in the **Tools | Board menu** (in this recipe it is the **Adafruit HUZZAH ESP8266**).
4. Select the serial port your board is connected to from the **Tools | Port menu** and then upload the code.

How it works...

The sketch creates a `servo` object called `myservo`, using the `servo` library, and attaches the `servo` object to GPIO 2 using the `attach(pin)` function. The syntax is `myservo.attach(2)`. This `servo` object is used to control the servo motor we have in our setup.

In the loop section of the project, a variable called `pos` is declared. It holds the position of the servo motor. To move the servo motor from 0 degrees to 180 degrees, a `for` loop is used. The `for` loop increments the position of the servo motor by 1 degree every 15 milliseconds. The `write()` function is used to move the servo to the new position every time the position is increased by 1 degree. The line that does that is `myservo.write(pos)`.

When the servo motor position reaches 180 degrees, the `for` loop ends and another `for` loop begins. The second `for` loop moves the servo motor position from 180 degrees back to the initial position of 0 degrees. It does this by decreasing the position by 1 degree every 15 milliseconds. When the position reaches 0 degrees, the whole process repeats.

You can calculate the time it takes the servo motor to move from 0 degrees to 180 degrees in our sketch, by multiplying the delay time by the number of iterations the `for` loop is going to make. So, in this case it is *15 milliseconds x 180 iterations = 2700* milliseconds or 2.7 seconds. The duration can be reduced by increasing the speed of the servo motor through altering the increment or decrement value in the `for` loops.

There's more...

Alter the sketch so that the servo motor moves faster than it is currently moving.

See also

Since you have mastered how to control outputs using the ESP8266, we will look at reading sensor input in the next recipe. You sure don't want to miss that.

Measuring data from a digital sensor

ESP8266 boards can be used to read and monitor data from both digital and analog sensors. To demonstrate this, we are going to measure data from a digital temperature and humidity sensor then display the data on the serial monitor.

Getting ready

In this tutorial, you will need an ESP8266 board, a USB cable, and a few other components, which include:

- DHT11 temperature/humidity sensor (`https://www.sparkfun.com/products/10167`)
- 10 kΩ resistor
- Breadboard
- Jumper wires

The DHT11 pin configuration is shown in the following diagram:

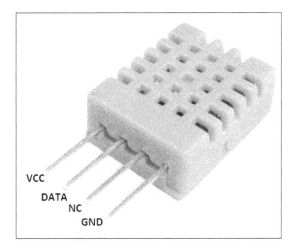

First mount the ESP8266 board and the DHT11 sensor onto the breadboard. Connect a 10 kΩ pull up resistor to the DHT11 data pin and connect the VCC pin and GND pin to the 3.3V pin and GND pin of the ESP8266 board, respectively. Finally, connect the data pin of the DHT11 to GPIO 2 of the ESP8266 board. Use jumper wires to do the connections.

The setup is shown in the following diagram:

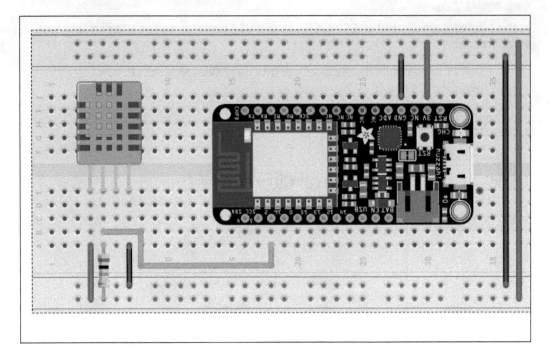

How to do it...

To measure temperature and humidity readings from the DHT11 sensor, we use the
`DHT` library from Adafruit. The library can be found at this link: `https://github.com/`
`adafruit/DHT-sensor-library`.

The library handles the reading of digital signals and the conversion of the digital signals
to more understandable formats, such as degrees Celsius and degrees Fahrenheit. It also
calculates the heat index from the readings. All this data can be accessed through the
use of some library functions such as `readTemperature()`, `readHumidity()`, and
`computeHeatIndex()`.

To demonstrate how to read digital sensors, we will obtain the humidity and temperature
measurements and display them on the serial monitor:

```
#include "DHT.h"

#define DHTPIN 2      // what digital pin we're connected to
#define DHTTYPE DHT11   // DHT 11

DHT dht(DHTPIN, DHTTYPE);
```

```
void setup() {
  Serial.begin(9600);
  dht.begin();
}

void loop() {
  // Wait a few seconds between measurements.
  delay(2000);

  // get humidity reading
  float h = dht.readHumidity();
  // get temperature reading in Celsius
  float t = dht.readTemperature();
  // get temperature reading in Fahrenheit
  float f = dht.readTemperature(true);

  // Check if any reads failed and exit early (to try again).
  if (isnan(h) || isnan(t) || isnan(f)) {
    Serial.println("Failed to read from DHT sensor!");
    return;
  }
  // display data on serial monitor
  Serial.print("Humidity: ");
  Serial.print(h);
  Serial.print(" %\t");
  Serial.print("Temperature: ");
  Serial.print(t);
  Serial.print(" *C ");
  Serial.print(f);
  Serial.println(" *F");
}
```

1. Copy the sketch and paste it in your Arduino IDE.
2. Check to ensure that the ESP8266 board is connected to the computer.
3. Select the board that you are using in the **Tools | Board menu** (in this recipe it is the **Adafruit HUZZAH ESP8266**).
4. Select the serial port your board is connected to from the **Tools | Port menu** and then upload the code.

How it works...

Our program includes the DHT library and creates a DHT object called dht. The DHT object defines the ESP8266 board GPIO pin where the DHT11 data pin is connected to and the type of DHT sensor that we are using. The GPIO pin is defined as 2 and the DHT type is defined as DHT11. The `Serial` interface and DTH object are then initialized in the setup section of the sketch.

The loop segment of the sketch starts with a 2-second delay that gives the sensor time to get readings. The humidity is then read using the `dht.readHumidity()` function and the temperature readings are acquired using the `dht.readTemperature()` function for readings in degrees Celsius, and `dht.readTemperature(true)` for readings in degrees Fahrenheit.

The sketch sends an alert if any of the readings are not obtained and tries again until it receives valid readings. If the obtained readings are alright, the program displays them on the serial monitor.

There's more...

Use the `computeHeatIndex()` function to get the heat index in Celsius and Fahrenheit. You can check the DHT tester example sketch that comes with the DHT library to get an idea of how to do that.

See also

Now that you have finished this recipe, proceed to the next one and learn how to display cool graphics and writing on an OLED screen using an ESP8266 board.

Controlling an OLED screen

To make your ESP8266 projects more interactive, you have to use more comprehensive ways of displaying data and information. This can be achieved with an OLED screen. Using it, you will be able to display writing and images, enabling one to know the current state of the project and to monitor its operations. In this recipe, you will learn how to control an OLED screen to display information on an ESP8266 project.

Getting ready

You will need an ESP8266 board, a USB cable, and a few other components, which include:

 ▶   Monochrome 1.3" 128x64 OLED graphics display (`https://www.adafruit.com/products/938`)

- ▶ Breadboard
- ▶ Jumper wires

The monochrome 1.3" 128x64 OLED graphics display pin configuration is shown in the following diagram:

The display has two interfaces that you can use to communicate with the ESP8266 board. They are i2C and SPI interface. When using the i2C interface, the pin connections should be as follows:

- ▶ GND of the display goes to GND of the ESP8266 board
- ▶ Vin of the display goes to 3V pin of the ESP8266 board
- ▶ Data pin of the display goes to the SDA pin (GPIO pin 4) of the ESP8266 board
- ▶ Clk pin of the display is connected to the SCL pin (GPIO pin 5) of the ESP8266 board
- ▶ RST pin can be connected to GPIO pin 2 of the ESP8266 board

The following figure shows how the connection should be:

The jumper pads shown by the red arrows should be closed or soldered together if you are using, the i2C interface. The setup will look the same as this:

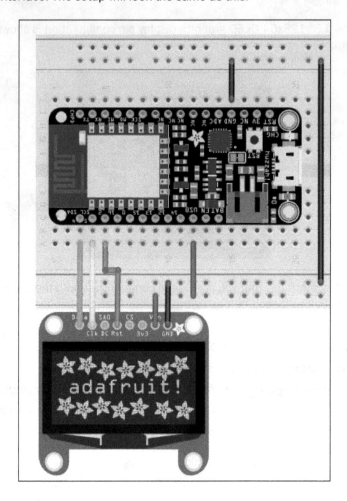

When using the SPI interface, the pin connections should be:

> ▸ GND pin of the display goes to the GND of the ESP8266 board
>
> ▸ Vin pin of the display goes to the 3V pin of the ESP8266 board
>
> ▸ Data pin of the display goes to the MOSI pin (GPIO pin 13) of the ESP8266 board
>
> ▸ Clk pin of the display goes to the SCK pin (GPIO pin 14) of the ESP8266 board
>
> ▸ D/C pin of the display is connected to GPIO pin 15 of the ESP8266 board
>
> ▸ RST of the display is connected to to GPIO pin 0 of the ESP8266 board
>
> ▸ CS of the display is connected to GPIO pin 16 of the ESP8266 board

The hardware connection will be as follows:

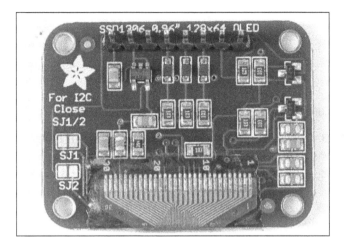

For the SPI connection, unsolder the **SJ1** and **SJ2** jumper pads. The setup will look the same as this:

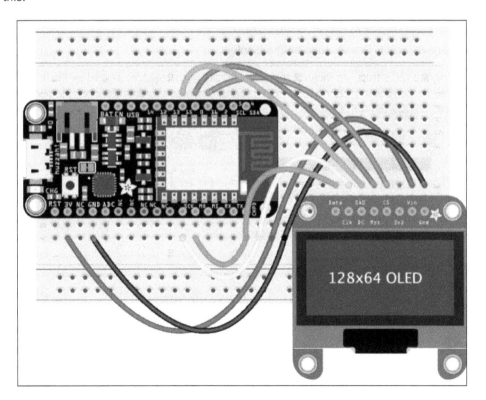

How to do it...

We use the `Adafruit SSD1306` library to control the OLED display. The library can be found at this link: `https://github.com/adafruit/Adafruit_SSD1306`.

We also need the `Adafruit GFX` library, which can be downloaded from this link: `https://github.com/adafruit/Adafruit-GFX-Library`.

Install the libraries on your Arduino IDE so that you can use them in your sketch.

The Adafruit SSD1306 has several functions that we use to display things on the OLED screen. Some of the common functions include:

- `clearDisplay()`: Clears the displays
- `setTextSize()`: Sets the size of text
- `setTextColor()`: Sets the text color
- `setCursor()`: Sets the cursor position
- `println()`: Prints text on the screen
- `drawBitmap()`: Draws a bitmap
- `invertDisplay()`: Inverts the shades
- `drawLine()`: Draws a line on the display

When using the i2C setup, we include the `wire` library in the sketch. The `wire` library comes preinstalled in the Arduino IDE so you won't need to download it. The i2C example code is as shown next. The sketch will display `Hello World` on the OLED screen:

```
#include <SPI.h>
#include <Wire.h>
#include <Adafruit_GFX.h>
#include <Adafruit_SSD1306.h>

#define OLED_RESET 2
Adafruit_SSD1306 display(OLED_RESET);

#define NUMFLAKES 10
#define XPOS 0
#define YPOS 1
#define DELTAY 2


#define LOGO16_GLCD_HEIGHT 16
#define LOGO16_GLCD_WIDTH  16

//#if (SSD1306_LCDHEIGHT != 64)
```

```
//#error("Height incorrect, please fix Adafruit_SSD1306.h!");
//#endif

void setup()   {
  Serial.begin(9600);

  // by default, we'll generate the high voltage from the 3.3v line
internally! (neat!)
  // initialize with the I2C addr 0x3D (for the 128x64)
  display.begin(SSD1306_SWITCHCAPVCC, 0x3D);

  // text display tests
  display.setTextSize(1);
  display.setTextColor(WHITE);
  display.setCursor(0,0);
  display.println("Hello, world!");

  delay(2000);
}

void loop() {

}
```

When using the SPI setup, we include the SPI library in the sketch. The SPI library comes preinstalled in the Arduino IDE so you won't have to download it. The SPI example code is as shown next. The sketch will display Hello World on the OLED screen:

```
#include <SPI.h>
#include <Wire.h>
#include <Adafruit_GFX.h>
#include <Adafruit_SSD1306.h>

// If using software SPI (the default case):
#define OLED_MOSI   13
#define OLED_CLK    14
#define OLED_DC     15
#define OLED_CS     16
#define OLED_RESET 0
Adafruit_SSD1306 display(OLED_MOSI, OLED_CLK, OLED_DC, OLED_RESET,
OLED_CS);

#define NUMFLAKES 10
#define XPOS 0
#define YPOS 1
```

```
#define DELTAY 2

#define LOGO16_GLCD_HEIGHT 16
#define LOGO16_GLCD_WIDTH  16


void setup()    {
  Serial.begin(9600);

  // by default, we'll generate the high voltage from the 3.3v line
internally! (neat!)
  display.begin(SSD1306_SWITCHCAPVCC);
  // init done

  // text display tests
  display.setTextSize(1);
  display.setTextColor(WHITE);
  display.setCursor(0,0);
  display.println("Hello, world!");
  delay(2000);
}


void loop() {

}
```

1. Copy the sketch and paste it in your Arduino IDE.

2. Check to ensure that the ESP8266 board is connected to the computer.

3. Select the board that you are using in the **Tools | Board menu** (in this recipe it is the **Adafruit HUZZAH ESP8266**).

4. Select the serial port your board is connected to from the **Tools | Port menu** and then upload the code.

How it works...

The i2C and SPI sketches work in the same way. The only difference is in the initialization of the communication interface. With the i2C sketch, we define the reset pin that we are using and then we create an `Adafruit_SSD1306` object with the defined reset pin. In this case, we are using GPIO 2 as the rest pin. This is done in the following lines of code:

```
#define OLED_RESET 2
Adafruit_SSD1306 display(OLED_RESET);
```

With the SPI sketch, we are required to define all the pins that are used in the SPI interface. Then we create an `Adafruit_SSD1306` object with all the defined SPI pins. This is done in the following lines of code:

```
#define OLED_MOSI   13

#define OLED_CLK    14

#define OLED_DC     15

#define OLED_CS     16

#define OLED_RESET  0

Adafruit_SSD1306 display(OLED_MOSI, OLED_CLK, OLED_DC, OLED_RESET, OLED_
CS);
```

The sketch then proceeds to define the coordinates, screen height and width, and other properties of the display.

In the setup section of the sketch, the Serial interface is initialized, as well as the `Adafruit_SSD1306` object, which is called `display` in our sketch. Once the OLED screen has been initialized, it displays the words `Hello world` and then goes to the loop section, which has no code.

There's more...

There are many other functions in the `Adafruit SSD1306` library that you can use to control the OLED display. Check them out in the examples provided in the `Adafruit SSD1306` library.

See also

Proceed to the next section to check the different issues that may arise when using the basic ESP8266 functions, and how to troubleshoot them.

Troubleshooting basic ESP8266 issues

You may run into some issues when using the ESP8266's basic functions. We will list some of the common problems that many people face and some ways of troubleshooting them.

The analog pin cannot measure high voltages

The analog pin of the ESP8266 can only measure voltages between 0V and 1V. If you have a sensor that outputs an analog signal that goes above that range, you will need to divide it. Otherwise, most of your readings from the analog pin will be 1023.

The best way of dividing the voltage is by using a voltage divider. It is easy to implement, since all you need is two resistors of the desired value and you are good to go.

Since most of the analog sensors you use with the ESP8266 board will output a signal of voltages between 0V and 3.3V, you will need a 1200 Ω resistor (R1) and a 470 Ω resistor (R2) to build a voltage divider.

You can read more on voltage dividers at this link: `https://learn.sparkfun.com/tutorials/voltage-dividers`.

The board stops working when things are connected to some pins

There are some pins on the ESP8266 board that are used for reset functions and for boot mode functions. Therefore, ensure that:

- CH_PD (EN) is always pulled high because it will disable the ESP8266 when it is pulled low
- RST is always pulled high because it will disable the ESP8266 when it is pulled low
- GPIO 0 is pulled high during power up/reset for the user program to run, otherwise the module will enter bootloader mode. It is normally pulled high by the red LED
- GPIO 2 is pulled high on power up/reset
- GPIO 15 is pulled low on power up/reset

The board keeps on crashing and resetting

This is mostly caused by a power failure. Always ensure you are powering the ESP8266 board from a 3.3V power supply if there is no on-board voltage regulator, or from a 5V power supply if there is an on-board voltage regulator.

The board produces gibberish on the serial monitor when I rest it

Those are ROM debug messages and the reason they appear as gibberish is because they are transmitter at a baud rate of 74880. They are usually corrupted and appear as a strange set of characters.

3

More ESP8266 Functions

In this chapter, we will cover:

- ▸ Discovering the advanced functions of the ESP8266
- ▸ Using libraries on the ESP8266
- ▸ Discovering the filesystem of the ESP8266
- ▸ Storing data in the ESP8266
- ▸ Discovering the **Over The Air** (**OTA**) update of the ESP8266
- ▸ Programming your ESP8266 OTA
- ▸ Troubleshooting basic ESP8266 issues

Introduction

The ESP8266 is a powerful standalone chip. In addition to general input and output functions, it offers wireless connectivity and provides you with a comprehensive filesystem for the storage of data among several other features. In this chapter, we will learn about the additional functions of the ESP8266 and how we can use them to improve our projects.

Discovering the advanced functions of the ESP8266

Using the ESP8266's advanced features, you can expand the scope of your projects. This recipe will look at some of the available additional features and their possible applications. Some of the additional features include:

- Wi-Fi connectivity
- Real-time clock
- Over the air update
- Low power management
- Working with files

Wi-Fi connectivity

The ESP8266 has a **Radio Frequency** (**RF**) transceiver with the following modules:

- 2.4 GHz transmitter and receiver
- Power management
- Real-time clock
- High-speed crystal oscillator and clock generator
- Regulators and bias

The RF transceiver supports 14 channels within the 2.4 GHz band in accordance to the `IEEE802.11bgn` standard. The channels range from 2.412 to 2.484 GHz. The RF receiver employs the use of **Automatic Gain Control** (**AGC**) integrated RF filters and DC offset compensation to accommodate the different signal channels.

A high-power CMOS amplifier drives the antenna. This enables the RF transmitter to achieve an average of +19 dBm transmission power in 802.11b transmission and +16 dBm in 802.11 n transmission.

Wi-Fi connectivity allows the ESP8266 to be used in **Internet of Things** (**IoT**) projects and to form mesh networks. When working as an IoT device, the ESP8266 module can either be a web server or a Wi-Fi client. When used as a web server, the ESP8266 hosts a web application or a web page that connected clients can access. As a Wi-Fi client, the ESP8266 connects to local or online servers and logs data to them or reads data from them.

Some IoT projects have many nodes that access the Internet. Generally, this would require very many routers since most of them support up to 32 nodes at a time. However, you can reduce the hardware requirements and cost of the project using the ESP mesh network. In this configuration, the different ESP8266 nodes form networks and forward packets to one node that transfers the packets to the router, as shown in the following figure:

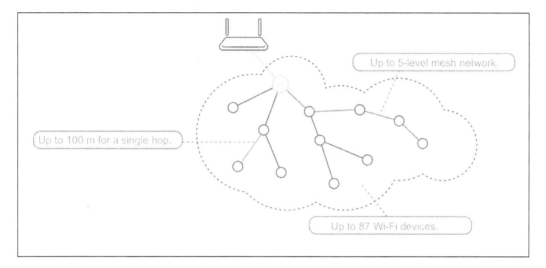

Real-Time Clock (RTC)

The ESP8266 module has an RTC counter derived from the internal clock. The RTC returns a 32-bit counter value that you can use to track time. Since it is a 32-bit counter, the RTC overflows after 7:45h. Therefore, if you want to implement a long term-working RTC, you must call the RTC function in your code once every 7 hours.

Over the air update

Over the air update refers to delivering new configurations or software to a remote device via wireless means. This is possible via the ESP8266 module. You can upload programs onto your ESP8266 from a remote location via Wi-Fi. We will look more into this later in this chapter.

Low power management

The ESP8266 module is designed for wearable electronics, IoT applications, and mobile devices. These devices require low power consumption since they usually have a limited supply of power, mostly from batteries. The ESP8266 module is designed with several power saving architectures that ensure proper low power management of your projects.

There are three modes of operation: deep sleep, sleep mode, and active mode. The deep sleep mode stops all other chip functions except for the RTC, which is left functional. This way you can continue tracking time even when in deep sleep mode.

The module draws only 60 μA when in deep sleep mode. It only has to be in active mode when transmitting or measuring data. By alternating between deep sleep and active modes you can run the ESP8266 module on a battery for years. This makes the ESP8266 module ideal for use in remote areas.

Working with files

When using NodeMCU firmware on the ESP8266, you can write and read scripts from the flash memory in file format instead of programming in raw memory locations. This is made possible by the **SPI flash filesystem** module, also known as **SPIFFS**. Using the filesystem you can easily manage your data on the module. This improves performance since data is easily cached, increasing the speed of access during operation. We will look more at working with files later on in this chapter.

See also

All those advanced features of the ESP8266 can be accessed when using the right libraries. Proceed to the next recipe to learn how to install and use those libraries so that you can access the additional functions on your ESP8266 board.

Using libraries on the ESP8266

In this recipe, we will learn how to use libraries with the ESP8266. The libraries will play a huge role in enabling us to access additional functions on our board. Apart from the Arduino core libraries that come with the Arduino IDE by default, we will need other libraries to explore the full potential of the ESP8266.

There are some ESP8266 libraries that get installed on the Arduino IDE automatically when you install the ESP8266 core, as explained in the first recipe of *Chapter 1, Configuring the ESP8266*. If you followed the instructions in *Chapter 1, Configuring the ESP8266*, you should already have those libraries on your Arduino IDE.

Some third-party ESP8266 libraries do not come with the ESP8266 core and you have to download them and install them on your Arduino. This is an easy procedure. You can use the Arduino library manager to search and download the library you need. To do that, open your Arduino IDE and navigate to **Sketch | Include Library | Manage Libraries...**:

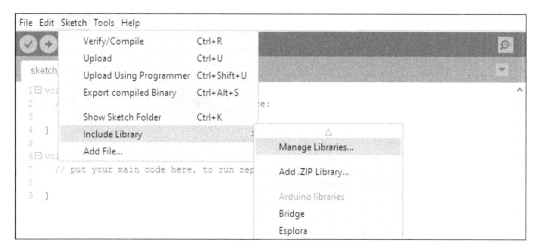

The **Library Manager** will open and list all the libraries that are already installed in your IDE, and also other libraries that are available for download. The **Library Manager** appears as shown in the following screenshot:

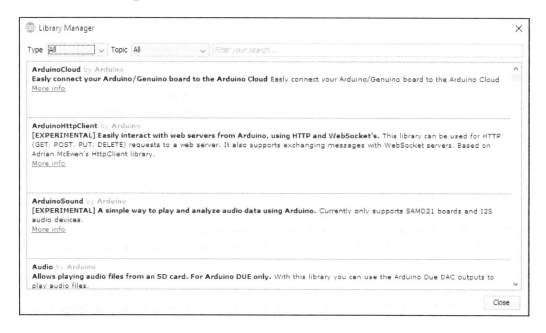

You can search for your ESP8266 library in the textbox `labeled filter your search` and install it or update it to the newest version.

If you opted to download the library in zip format, just open the Arduino IDE, then go to **Sketch | Include Library | Add .ZIP Library...** as shown in the following screenshot, and then navigate to where your library ZIP file is and select it:

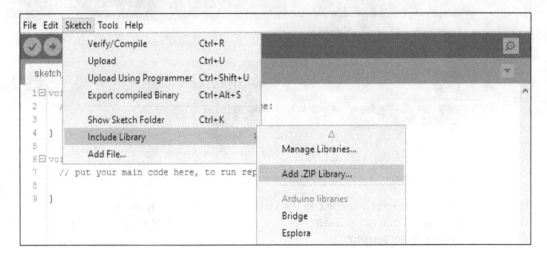

For more information about installing `Arduino` libraries, you can visit this link: `https://www.arduino.cc/en/Guide/Libraries`.

We will look at a Wi-Fi web server example to demonstrate how to use ESP8266 libraries.

Getting ready

The hardware we need for this tutorial is the ESP8266 board and a USB cable. Connect your ESP8266 board to your computer.

How to do it...

We are going to create a simple Wi-Fi web server using the `ESP8266WiFi` library. The web server will send a `Hello World!` reply whenever a client connects to it. To use a library, you have to add it at the beginning of the sketch using the `#include` pre-processor directive.

In our case, we used the line `#include<ESP8266WiFi.h>` to include the library. The library has many functions that we will use to connect to the Wi-Fi network, monitor whether clients have connected to our server, and also send data to the clients. This gives us full control of our simple Wi-Fi server:

```
#include <ESP8266WiFi.h>

// Create an instance of the server
// specify the port to listen on as an argument
```

```
WiFiServer server(80);

const char* ssid = "your-ssid";
const char* password = "your-password";

void setup() {
  Serial.begin(115200);
  delay(10);

  // Connect to WiFi network
  Serial.print("Connecting to WiFi hotspot");
  WiFi.begin(ssid, password);

  while (WiFi.status() != WL_CONNECTED) {
    delay(500);
    Serial.print(".");
  }
  Serial.println("");
  Serial.println("WiFi connected");

  // Start the server
  server.begin();
  Serial.println("Server started");

  // Print the IP address
  Serial.println(WiFi.localIP());
}

void loop() {
  // Check if a client has connected
  WiFiClient client = server.available();
  if (!client) {
    return;
  }
  // Wait until the client sends some data
  while(!client.available()){
    delay(1);
  }
  // Prepare the response
  String s = "HTTP/1.1 200 OK\r\nContent-Type: text/html\r\n\r\
n<!DOCTYPE HTML>\r\n<html>\r\nHello World! </html>\n";
  // Send the response to the client
  client.print(s);
  delay(1);
  Serial.println("Client disonnected");
}
```

1. Copy the sketch and paste it in your Arduino IDE.

2. Check to ensure that the ESP8266 board is connected. Select the board that you are using in the **Tools | Board menu** (in this case it is the **Adafruit HUZZAH ESP8266**).

3. Select the serial port your board is connected to from the **Tools | Port menu** and then upload the code. Open the serial monitor to check the debug messages.

How it works...

The sketch includes the `ESP8266WiFi` library and then creates a server instance from the library using the line `WiFiServer server(80);`. The server instance specifies the port that should be listened to, which in this case is port `80`. The SSID and password of the Wi-Fi hotspot the ESP8266 board will connect to are then defined.

The setup section of the sketch initializes the serial interface and then connects the ESP8266 to the Wi-Fi hotspot using the `WiFi.begin()` function. When the connection to the Wi-Fi network is successfully established, the server is started using the `server.begin()` function and the IP address of the ESP8266 is obtained using the `WiFi.localIP()` function. The IP address is then printed onto the serial monitor.

In the loop section, the program checks whether there is a client connected to the ESP8266 server using the `server.available()` function. If there is a connected client, the ESP8266 listens to it with the `client.available()` function. If there is data from the client, the server sends the response `Hello World!` to the client using the `client.print()` function and then disconnects the client.

To test the program, connect your phone or computer to the same Wi-Fi network as the ESP8266 module. Use a web browser to access the server via this URI: `http://Server_IP/`. Replace `Server_IP` in the URI with the IP address of your ESP8266 board. If you do not know the IP address of your ESP8266 module, check the serial monitor. It will be displayed when the ESP8266 successfully connects to the Wi-Fi network and starts the server. When you successfully access that URI, you will see `Hello World!` displayed on your web browser.

There's more...

Try some of the examples in the `ESP8266` libraries that are available on your Arduino IDE and check how the libraries are used.

See also

The ESP8266 has an elaborate filesystem that you can use to store data easily. We are going to look at that filesystem in the next recipe.

Discovering the filesystem of the ESP8266

One of the advantages that the ESP8266 has over most other IoT boards is the ability to store files in its flash memory. You can store configuration files, sketch data, and even content for a web server in the flash memory. In this recipe, we will learn about the architecture of the flash memory and the filesystem as a whole. This will enable us to know how to properly store and retrieve data in and from it.

Flash memory

In most microcontroller architectures, the flash memory is used as the main storage media. Configuration files, firmware, and application data are all stored in it. The flash memory allows random writes and reads and it can only be erased in blocks at a time. Moreover, it can only be erased a specific number of times (between 100,000 and 1,000,000 times) before it fails.

In the Arduino environment, the flash memory is laid out in a specific way, as shown in the following screenshot. This is to ensure proper memory use and maintenance:

```
|---------------|-------|---------------|--|--|--|--|--|
       ^              ^          ^             ^    ^
   Sketch      OTA update   File system    EEPROM  WiFi config (SDK)
```

This memory architecture is used in the ESP8266 and the memory allocation we are going to be concerned with is the one set aside for the filesystem. The size of the filesystem usually varies, depending on the size of the flash memory. Different ESP8266 boards will not have the same flash memory size. As such, the Arduino IDE sets the available filesystem size depending on the board you select. The filesystem sizes are as follows:

Board	Flash chip size, bytes	Filesystem size, bytes
Generic module	512 k	64 k, 128 k
Generic module	1 M	64 k, 128 k, 256 k, 512 k
Generic module	2 M	1 M
Generic module	4 M	3M
Adafruit HUZZAH	4 M	1M, 3 M
ESPresso Lite 1.0	4 M	1 M, 3 M
ESPresso Lite 2.0	4 M	1 M, 3 M
NodeMCU 0.9	4 M	1 M, 3 M
NodeMCU 1.0	4 M	1 M, 3 M
Olimex MOD-WIFI-ESP8266(-DEV)	2 M	1 M

Board	Flash chip size, bytes	Filesystem size, bytes
SparkFun Thing	512 k	64 k
SweetPea ESP-210	4 M	1 M, 3 M
WeMos D1 & D1 mini	4 M	1 M, 3 M
ESPDuino	4 M	1 M, 3 M

Filesystem memory management

Working with flash memory is a difficult task, unless you are using SPIFFS. It is a module that enables access to and storage of data in the flash memory in file format.

SPIFFS partitions the available memory into pages to create a filesystem. The pages are usually small enough to be cached in the RAM. The default size is 256 bytes. The pages contain file contents and index information.

SPIFFS optimizes the writes to conserve flash memory and reduce the number of erases performed. When small writes are required, they are first written in a buffer in the RAM. When the buffer fills up, the data is written to the flash memory and a new page is started. In case there is data in the middle of a page that needs to be changed, the changes are copied to a new page and the former page is deleted.

Setting up the ESP8266FS tool

The ESP8266FS is a tool that enables the Arduino IDE to upload files onto the ESP8266 flash filesystem. When the tool is properly installed, a menu item appears in the **Tools** menu of the Arduino IDE. You can use that menu item to do the uploading.

To successfully set up the ESP8266FS tool, follow these steps:

1. Download the ESP8266FS tool from this link: https://github.com/esp8266/arduino-esp8266fs-plugin/releases/download/0.2.0/ESP8266FS-0.2.0.zip.

2. Create a Tools directory in your Arduino sketchbook directory.

3. Unzip the ESP8266FS tool into the Tools directory. The path will look like this: My Documents/Arduino/tools/ESP8266FS/tool/esp8266fs.jar.

4. Restart the Arduino IDE.

5. Create a new sketch or open one that already exists.

6. Click on the Sketch | Show sketch folder.

7. Create a directory called data.

8. Ensure that all the files you are intending to save in the filesystem are in the `data` directory.

9. Select the board and port and close the serial monitor if it is open.

10. Click on **Tools | ESP8266 Sketch Data Upload**.

11. The IDE will start uploading the files in the `data` directory to the ESP8266 flash filesystem.

Once the upload is complete, the status bar on the IDE will display **SPIFFS image Uploaded**.

The library that supports SPIFFS is called `FS`. Therefore, if you want to store some files in the ESP8266 flash filesystem, add the `#include <FS.h>` pre-processor directive in your sketch to include the `FS` library.

Some of the common functions in the `FS` library you will be using include:

- `Filesystem` object:
    - `begin()`: Mounts the SPIFFS filesystem
    - `format()`: Formats the filesystem
    - `end()`: Unmounts the SPIFFS filesystem
    - `open(path, mode)`: Opens a file
    - `exists(path)`: Returns true if there is a file with the provided path
    - `openDir(path)`: Opens a directory whose path is provided
    - `rename(pathfrom, pathTo)`: Renames files
    - `remove(path)`: Deletes the provided path

- `Directory` object:
    - `next()`: Returns true if there are files to scroll over
    - `fileName()`: Returns the filename of the file that is being pointed to
    - `openfile(mode)`: Opens the file

- `File` object:
    - `seek(offset, mode)`: Returns true if the position was set successfully
    - `position()`: Returns current position in the file in bytes
    - `size()`: Returns the size of the file in bytes
    - `name()`: Returns the filename
    - `close()`: Closes the file

Using these functions, you will be able to realize the full potential of the ESP8266 filesystem. It comes in handy, especially in web server applications for storage of web pages and other web-related resources. You can also use it to save sensor data or important configurations.

There's more...

Read more on the SPIFFS library and functions at this link: `https://github.com/esp8266/Arduino/blob/master/doc/filesystem.md`.

See also

Now that you have understood the ESP8266 filesystem, you should apply it in your sketches. Proceed to the next recipe and learn how to use the filesystem to store data in your projects.

Storing data in the ESP8266 filesystem

In this recipe, we will be looking at how to store data in the ESP8266 filesystem. To do that, we will store data measured by a sensor in the flash filesystem. This will get you started on using SPIFFS to store files on your ESP8266, and give you an idea of how you can use the filesystem in your projects.

Getting ready

Make sure you have the following components before proceeding:

- ESP8266 board
- USB cable
- DHT11 temperature/humidity sensor (`https://www.sparkfun.com/products/10167`)
- 10 kΩ resistor (`https://www.sparkfun.com/products/8374`)
- Breadboard
- Jumper wires

Start by mounting the ESP8266 board and the DHT11 sensor onto the breadboard. Connect a **10** kΩ pull up resistor to the DHT11 data pin and connect the VCC pin and GND pin to the 3V pin and **GND** pin of the ESP8266 board, respectively. Finally, connect the data pin of the DHT11 to GPIO **2** of the ESP8266 board. Use jumper wires to do the connections. The complete setup will look as shown in the following diagram:

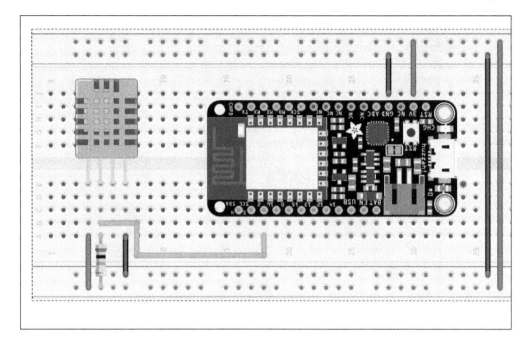

How to do it...

Refer to the following steps:

1. The first thing we do is create a text file called `log.txt` that will hold the sensor data. There are two ways you can go about it. You can use the Arduino IDE to upload the text file to the ESP8266 filesystem or you can create it using our sketch.

2. To upload it using the Arduino IDE, you create a directory called `data` in the `Sketch` folder. You can access the `Sketch` folder as shown in the following screenshot:

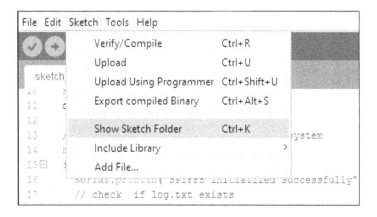

3. Then, create the `log.txt` file in the `data` folder:

4. With the `log.txt` file successfully created, go back to the Arduino IDE. Select the ESP8266 board you are using and the correct serial port from the tools menu, and click on **Tools | ESP8266 Sketch Data Upload**.

The `log.txt` file will be uploaded to the ESP8266 board filesystem.

 This only works if you have installed the ESP8266FS tool as explained in the previous recipe.

We designed our sketch such that if the `log.txt` file does not exist, it is created. This is handled using the append mode for file access, which automatically creates the file if the sketch does not find it.

To demonstrate saving data to the filesystem, we sampled temperature data from a `DHT11` sensor and saved the data into a txt file every 10 seconds:

```
#include "FS.h"
#include "DHT.h"

#define DHTPIN 2 // the digital pin we are connected to
#define DHTTYPE DHT11 // define type of DHT sensor we are using

DHT dht(DHTPIN, DHTTYPE); // create DHT object
```

```
void setup() {
  Serial.begin(115200);
  dht.begin(); // intialize DHT object

  // always use this to "mount" the filesystem
  bool ok = SPIFFS.begin();
  if (ok) {
    Serial.println("SPIFFS initialized successfully");
  }
  else{
    Serial.println("SPIFFS intialization error");
  }


}

void loop() {
  // Wait a few seconds between measurements.
  delay(2000);
  //read temperature as Celcius
  float t = dht.readTemperature();

  // Check if any reads failed and exit early (to try again).
  if (isnan(t)) {
    Serial.println("Failed to read from DHT sensor!");
    return;
  }
  //open log.txt file
  File f = SPIFFS.open("/log.txt", "a");
  if (!f) {
    Serial.println("file open failed");
  }
  // save temperature reading
  f.print(t);
  f.println("deg C");
  //close file
  f.close();
  delay(8000);
}
```

5. Copy the sketch and paste it in your Arduino IDE.

6. Check to ensure that the ESP8266 board is connected.

7. Select the board that you are using in the **Tools | Board menu** (in this case it is the **Adafruit HUZZAH ESP8266**).

8. Select the serial port your board is connected to from the **Tools | Port menu** and then upload the code.

How it works...

The program includes both the DHT and FS libraries, which provide functions for obtaining data from the sensor and for managing the filesystem, respectively. The DHT object is created. It defines the ESP8266 board GPIO pin that the DHT11 data pin is connected to and the type of DHT sensor that we are using. The GPIO pin is defined as 2 and the DHT type is defined as DHT11.

In the setup section of the sketch, the serial interface, the DHT object, and SPIFFS are initialized. A notification message is displayed on the serial monitor to show whether SPIFFS has been successfully initialized or not.

The loop segment of the sketch starts with a two second delay that gives the sensor time to get readings. The temperature readings are acquired using the dht.readTemperature() function for readings in degrees Celsius. An alert is displayed on the serial monitor if the temperature reading is not obtained and the sensor is read again until valid readings are acquired.

If the obtained readings are OK, the program opens the log.txt file in append mode. If the file does not exist, it is created. This is accomplished in the SPIFFS.open("/log.txt", "a") function. If the file is not successfully opened, an alert is displayed on the serial monitor, otherwise the temperature is saved in the txt file. The file is then closed, followed by an eight second delay.

There's more...

Write a sketch that reads the sensor data that has been saved in the log.txt file in the ESP8266 filesystem.

See also

Programming your ESP8266 board using a USB cable is easy. However, it is not practical when your board is located in a remote area. Wouldn't it be great if you could upload your sketches to your board wirelessly? Well, that is possible and the next recipe will show you exactly how to do it.

Discovering the over the air update of the ESP8266 (OTA)

In this recipe, we are going to learn about the OTA update functionality of the ESP8266. This will enable you to understand how OTA works and how you can apply it in your projects. With this knowledge, you will be able to easily program your ESP8266 boards that are situated in remote locations.

Over the air

An OTA update involves loading firmware to an ESP8266 module via Wi-Fi, instead of using a serial port. This is a very important feature that ensures the delivery of firmware updates to ESP8266 boards in remote areas. You can do OTA using an Arduino IDE, a HTTP server, or a web browser.

The Arduino IDE is usually used for the software development phase. The other two are more important after deployment. They allow you to provide modules with application updates either automatically with a HTTP server or manually using a web browser.

Initial firmware is uploaded onto the ESP8266 board via a serial port. Once the OTA routines are implemented correctly, all other uploads can be done through OTA updates.

There are several considerations that you need to make when implementing OTA routines. The first one is that there are no security measures that prevent the OTA process from being hacked. Therefore, it is up to you to ensure that only updates from legitimate sources are accepted.

Moreover, OTA updates interrupt the normal running of the previous code, once the new code is executed. Therefore, it is important for you to make sure that previous code is terminated and restarted safely during and after the update. In some cases, the ESP8266 may be controlling some vital processes that would cause harm or damage if not stopped in the right manner. Examples of such procedures include an irrigation system. If OTA updates are not implemented in the right way, a valve may be left open accidentally and cause flooding.

Security

To reduce cases of hacking when using OTA, you can protect your updates with passwords or accept updates only from a specified OTA port. The `Arduino OTA` library provides a check functionality that improves security using these functions:

```
void setPort(uint16_t port);
void setHostname(const char* hostname);
void setPassword(const char* password);
```

There is also an inbuilt protection functionality that `Arduino OTA` and `espota.py` use to authenticate uploads. It is known as Digest-MD5. The integrity of the data is checked using the MD5 checksum. No coding is needed on your part to implement this.

In addition to the inbuilt protection and library functions, you can also add your own safeguards. Some of them include accepting updates according to a specific schedule or limiting OTA triggering to some specific actions, such as the user pressing a dedicated update button.

Safety

When an OTA upload is taking place, some ESP resources, such as bandwidth, are used. The module then restarts after the update. These things might affect the current operations of your module and have adverse effects on the controlled processes. Therefore, you should test how the updates affect the functionality of your new and existing sketch before deploying your module to the field.

The best practice is putting the controlled processes or equipment in a safe state before beginning the update. There are several functions in the `Arduino OTA` library that handle the functionality of your application, during the different stages of OTA or in case of an OTA error. They include:

```
void onStart(OTA_CALLBACK(fn));
void onEnd(OTA_CALLBACK(fn));
void onProgress(OTA_CALLBACK_PROGRESS(fn));
void onError(OTA_CALLBACK_ERROR (fn));
```

You can use these functions to keep your equipment and processes safe during updates.

 When using OTA, ensure that the flash chip size is double the size of the sketch.

See also

If OTA fascinates you and you would like to try it out in your projects, proceed to the next recipe and learn how to implement it.

Programming your ESP8266 OTA

This recipe is going to demonstrate how to upload a sketch to an ESP8266 board over the air. We will upload a simple LED blinking program onto our ESP8266 board using OTA and implement some OTA routines to ensure safety during the upload.

Getting ready

In this recipe, we will need the following components:

- ► ESP8266 board
- ► USB cable
- ► LED (`https://www.sparkfun.com/products/528`)
- ► 220 Ω resistor (`https://www.sparkfun.com/products/10969`)
- ► Breadboard
- ► Jumper wires

Start by mounting the LED onto the breadboard. Connect one end of the 220 Ω resistor to the positive leg of the LED (the positive leg of an LED is usually the taller one of the two legs). Connect the other end of the resistor to pin **5** of the ESP8266 board. Then use a jumper wire to connect the negative leg of the LED to the **GND** pin of the ESP8266.

The setup is shown in the following diagram:

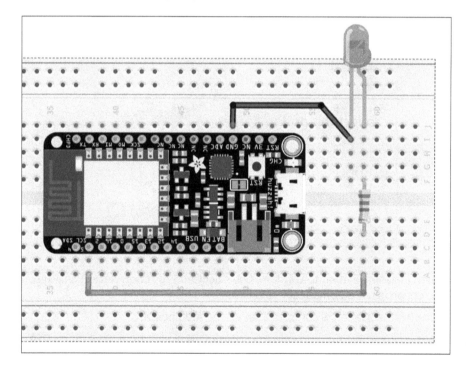

How to do it...

The first thing you need to do is set up your Arduino IDE for OTA updates. You will require the following things:

- ▶ Python 2.7 (`https://www.python.org/`).
- ▶ Arduino core for ESP8266 (`https://github.com/esp8266/Arduino#installing-with-boards-manager`)
- ▶ Arduino IDE 1.6.7 or later versions
- ▶ A Wi-Fi network to which your computer and ESP8266 board will be connected

Since we already tackled how to install the Arduino core for ESP8266 in the first chapter, we will only look at installing Python 2.7 and how to proceed from there. Therefore, download the Python 2.7 setup and run it. The setup window appears as shown in the following screenshot:

1. When installing Python to Windows OS, select `Add python.exe to Path`, since the option is not automatically selected.

2. Once the installation is complete, set up the Arduino IDE.

3. Open your Arduino IDE and select the **BasicOTA** sketch from the **ArduinoOTA** examples library:

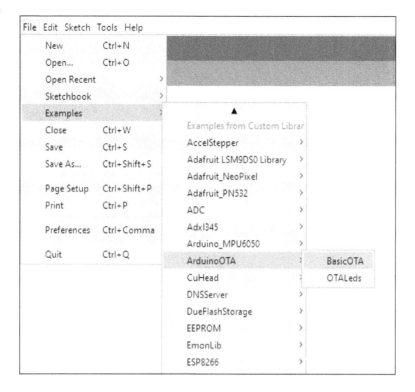

4. Add your Wi-Fi network credentials to the sketch by editing the following lines:

```
const char* ssid = "..........";
const char* password = "..........";
```

5. Type in the ssid and password of your Wi-Fi network between the double quotation marks.

6. Check to ensure that the ESP8266 board is connected to the computer.

7. Select the board that you are using in the **Tools | Board menu** (in this recipe it is the **Adafruit HUZZAH ESP8266**).

8. Select the serial port your board is connected to from the **Tools | Port menu** and then upload the sketch.

9. Open the serial monitor. You should see the IP address of your ESP8266 board displayed. That shows that your board has successfully connected to the Wi-Fi network.

10. Wait for a short while and check under **Tools | Ports | Network ports** on your Arduino IDE. You should see a new OTA port:

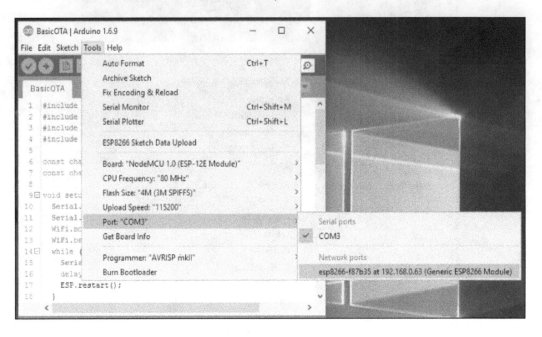

11. If the OTA port does not appear, restart the Arduino IDE. If it still doesn't appear, check your firewall settings.

12. Select the **OTA** port. The **Tools** menu will then appear as follows:

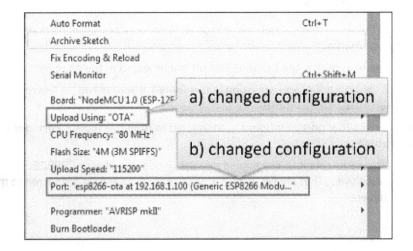

13. Your ESP8266 board is ready to receive OTA updates. Now you can upload the LED blinking program with OTA routines. The code is as follows:

```
#include <ESP8266WiFi.h>
#include <ESP8266mDNS.h>
#include <WiFiUdp.h>
#include <ArduinoOTA.h>

const char* ssid = "yourssid";
const char* password = "yourpassword";

void setup() {
  Serial.begin(115200);// initialize serial interface
  pinMode(2, OUTPUT); // set GPIO 2 as output
  Serial.println("Booting");
  // set ESP8266 mode
  WiFi.mode(WIFI_STA);
  // connect to Wi-Fi network
  WiFi.begin(ssid, password);
  while (WiFi.waitForConnectResult() != WL_CONNECTED) {
    Serial.println("Connection Failed! Rebooting...");
    delay(5000);
    ESP.restart();
  }

  // routines that will take place during OTA update
  // start routine
  ArduinoOTA.onStart([]() {
    Serial.println("Start"); //display start on serial monitor
    digitalWrite(2, LOW); // set GPIO 2 state LOW
  });
  // end routine
  ArduinoOTA.onEnd([]() {
    Serial.println("\nEnd");//display start on serial monitor
    digitalWrite(2, HIGH); // set GPIO 2 state HIGH
  });
  // progress routine
  // show the OTA update progress on serial monitor
  ArduinoOTA.onProgress([](unsigned int progress, unsigned int
total) {
    Serial.printf("Progress: %u%%\r", (progress / (total / 100)));
  });
  // error routines
  ArduinoOTA.onError([](ota_error_t error) {
    Serial.printf("Error[%u]: ", error);
```

```
          if (error == OTA_AUTH_ERROR) Serial.println("Auth Failed");
          else if (error == OTA_BEGIN_ERROR) Serial.println("Begin
          Failed");
          else if (error == OTA_CONNECT_ERROR) Serial.println("Connect
          Failed");
          else if (error == OTA_RECEIVE_ERROR) Serial.println("Receive
          Failed");
          else if (error == OTA_END_ERROR) Serial.println("End Failed");
        });

        // setup OTA server
        ArduinoOTA.begin();
        Serial.println("Ready");

        // print ESP8266 IP address
        Serial.print("IP address: ");
        Serial.println(WiFi.localIP());
      }

    void loop() {
      ArduinoOTA.handle();
      // blink program
      digitalWrite(2, HIGH);
      delay(500);
      digitalWrite(2, LOW);
      delay(500);
    }
```

14. Copy the sketch and paste it in your Arduino IDE. Type in the SSID and password of your Wi-Fi network at the beginning of the sketch as you did with the **BasicOTA** sketch.

15. Select the board that you are using in the **Tools | Board menu** (in this recipe it is the **Adafruit HUZZAH ESP8266**).

16. Make sure you have selected the OTA port in the **Tools | Port menu** and then upload the code.

How it works...

The sketch includes all the required libraries and then saves the SSID and password in constant variables. In the setup section, the serial interface is initialized and GPIO **2** set as an output. The ESP8266 module is set in STA mode and then connects to the Wi-Fi network using the SSID and password that you provided.

OTA routines are then set. When the OTA update is started the GPIO 2 state is set to low, turning off the LED, and when it ends the GPIO 2 is set to high, turning on the LED. When the OTA update is in progress, the progress routine displays the status of the update on the serial monitor. If there is an error, the error routine displays the kind of error that has been experienced on the serial monitor.

Once the OTA update is done, the OTA server is set up and the ESP8266 IP address is displayed on the serial monitor. The loop section contains the OTA handle, which implements the LED blinking sketch that is in the loop.

The LED blinking program turns on the LED, delays for 500 milliseconds, turns the LED off, and then delays for a further 500 milliseconds before turning on the LED again.

There's more...

Explore the different ways of improving OTA update security on your sketch. You can use these functions to do that:

```
void setPort(uint16_t port);
void setHostname(const char* hostname);
void setPassword(const char* password);
```

Check the `OTALed.ino` example in the **ArduinoOTA** library for insight into how to do it.

See also

Check the next recipe for some of the common errors you may encounter when using advanced ESP8266 functions and how to troubleshoot them.

Troubleshooting basic ESP8266 issues

You may run into some issues when using the ESP8266's advanced functions. We will list some of the common problems that many people face and some ways of troubleshooting them.

The Arduino IDE does not show the OTA port

There are three scenarios in which this could happen:

- The ESP8266 board has not successfully connected to the Wi-Fi network
- The firewall is blocking your Arduino IDE from accessing the OTA port
- The Arduino IDE needs a restart

To solve the issue, start by restarting the Arduino IDE. If that does not work, troubleshoot the first scenario by checking whether the ESP8266 IP address was displayed on the serial monitor after uploading the OTA code. If it wasn't, check the Wi-Fi credentials in your sketch and make sure they are correct.

If all that does not work, try changing the firewall settings on your computer, so that it allows the Arduino IDE to access Wi-Fi network features.

The library no longer compiles

This sometimes happens after updating libraries. It can be solved by deleting the library in question and reinstalling the ESP8266 core using the board manager.

The txt file in my filesystem only holds one value even after saving several values in it

This is caused by the use of the wrong file access mode of the SPIFFS open function. When you use w or w+ as the file access mode, it clears the file every time you open it. That way the only available data in the file is what you saved last. Instead of using w or w+ file access modes, you should use a or a+. In these file modes, any data that is added to the file is appended, at the end of file, to the already existing data. Therefore, the function should look like this:

```
SPIFFS.open("/filename.txt", "a");
```

Instead of like this:

```
SPIFFS.open("/filename.txt", "w");
```

4

Using MicroPython on the ESP8266

In this chapter, we will cover:

- ▶ Introduction to MicroPython on the ESP8266
- ▶ Discovering the MicroPython language
- ▶ Running MicroPython on the ESP8266
- ▶ Controlling pins using MicroPython
- ▶ Reading data from a sensor using MicroPython
- ▶ Sending data to the cloud using MicroPython
- ▶ Troubleshooting common MicroPython issues

Introduction

In this chapter, we will learn how to configure and control an ESP8266 board using MicroPython. MicroPython is one of the many programming languages that we can use to program the ESP8266 module. It is a lean and fast version of the Python 3 programming language and has several advantages over traditional programming languages such as C and C++.

The first advantage is that Python is a very well-known programming language. Therefore, there are numerous resources and libraries available online. This comes in handy when working on your open source projects.

Writing and debugging code using MicroPython is faster than when using other programming languages. In fact, it is about five times faster than C++. It is also easier to program using MicroPython because it is expressive and versatile, which makes it great for beginners.

MicroPython is great for low power applications, as the chip consumes less power when booting with MicroPython. This makes devices very energy efficient and ideal for low power applications that required a long battery life.

Introduction to MicroPython on the ESP8266

One way you can get the most out of your ESP8266 is by using MicroPython. Also, the ESP8266 module is one of the best platforms on which to learn how to use MicroPython. This is because the ESP8266 provides simple GPIO pin control functions as well as wireless functionality, allowing you to test all aspects of the MicroPython programming language.

MicroPython & the ESP8266

MicroPython is an implementation of the Python 3 programming language that features a small portion of the Python standard library. It is designed to be lightweight and more efficient in order to run effectively on microcontrollers and other environments that have limited processing power and storage space.

Despite being lean, MicroPython has many advanced features, such as an interactive prompt, exception handling, arbitrary precision integers, list comprehension, and closures, among several other features. You can access and run all these features in devices that have less than 256 KB flash memory and 16 KB RAM. This way you are able maximize the functionality of your projects.

MicroPython is designed to be compatible with normal Python as much as possible. It has a complete Python compiler and runtime, and provides an interactive prompt known as REPL that executes commands immediately. MicroPython also provides the ability to import and execute scripts saved in the built-in filesystem. If you know how to code using Python, you will automatically be able to code using MicroPython.

MicroPython is designed to support many microcontroller architectures. Some of the most notable architectures include Xtensa, ARM, x86m ARM Thumb, and x86-64. This makes it easy to move firmware between different microcontrollers without having to overhaul the whole code.

The ESP8266 chip is popular in the open source development industry. There are many development boards from different manufacturers that use the ESP8266 chip. MicroPython has been designed to provide a generic port that can run on most of those boards, with as few limitations as possible. The port is based on the Adafruit Feather HUZZAH board. When using other ESP8266 boards, make sure you check their schematics and datasheets so that you can identify the differences between them and the Adafruit Feather HUZZAH board. That way, you can accommodate the differences in your code.

The MicroPython ESP8266 port follows several design rules and considerations. They include:

- GPIO pin numbers are based on the ESP8266 chip pin numbering. To make sure your board conforms to this, get its pin diagram and check how its pins correspond to the ESP8266 pin numbers. If the pin numbers do not match, note the differences so that you can account for them when writing your code.

- All relevant pins that are dedicated to a specific function, such as the pins connected to the SPI flash memory, are supported by MicroPython. Some boards will not expose these pins. That is why you should confirm your board's pin configuration before proceeding.

- Not all boards have external pins or internal connectivity that supports the ESP8266 deep sleep mode.

When programming the ESP8266 using MicroPython, it is important for you to remember that the chip has scarce runtime resources. For instance, you will only have about 32 KB of RAM to use. Therefore, avoid creating big container objects and buffers. Moreover, since there is no software that manages the on-chip resources, it is up to you to make sure that you close any open files as soon as you finish using them.

The booting process involves several actions. It starts by executing the `boot.py` script from the internal modules that had been frozen. The filesystem gets mounted in the flash memory. If the filesystem is not available, the MicroPython ESP8266 port does the initial setup of the module and creates the filesystem.

After the successful mounting of the filesystem, the `boot.py` script is executed. This starts the WebREPL daemon. Once that is done, the `main.py` script gets executed from the filesystem. This file starts the user application every time the ESP8266 boots up.

The ESP8266 has a **Real-Time Clock** (**RTC**). However, it is, limited to tracking time up to 7:45 hrs. If you want to use the RTC to track long periods of time, you will have to call the `time()` and `localtime()` functions every seven hours.

See also

Now that you have an idea of what MicroPython is you should be itching to create your first program with it. Proceed to the next recipe to learn a few of the basics you need when programming using MicroPython.

Discovering the MicroPython language

In this recipe, we will look at some of the basics of MicroPython. This will give you a clear understanding of the structure and different features of the MicroPython programming environment.

Discovering MicroPython

MicroPython is basically Python with some additional libraries to control boards. It comes with an interactive prompt known as the MicroPython Interactive Interpreter mode or **Read-Eval-Print-Loop** (**REPL**). REPL runs and imports scripts from the built-in filesystem. It also comes with several features that enhance your coding experience. These features include:

- Auto-indent
- Auto-completion
- Interrupting a running program
- Paste mode
- Soft reset
- Special variable
- Raw mode

Auto-indent

The auto-indent feature keeps your programs neat. When you type a statement that ends with a colon (`:`), such as `for` or `while` loops or `if` statements, the prompt changes to three dots (...) and the cursor gets indented by four spaces. The next lines after that continue on the same level when you press the *Enter* key.

The *backspace* key is used to undo indentations. One press removes one level of indentation. When you press *Enter* while the cursor is back at the beginning, the code you have entered gets executed.

The following screenshot shows what happens when you enter a `for` statement. The underscore represents the cursor:

```
>>> for i in range(3):
...     _
```

If you include an `if` statement in the code, another level of indentation is added, as shown here:

```
>>> for i in range(30):
...     if i > 3:
...         _
```

When you type break, then press the *Enter* key once and *backspace* key once, the code will look the same as this:

```
>>> for i in range(30):
...       if i > 3:
...             break
...       _
```

Type print(i), then press the *Enter* key once to go to a new line, then the *backspace* key once to remove one level of indentation, and then press the *Enter* key again to execute the program. The prompt will look the same as this:

```
>>> for i in range(30):
...       if i > 3:
...             break
...       print(i)
...
0
1
2
3
>>>
```

You can also execute the program by pressing the *Enter* key three times. This is because auto-indent is usually not applied if the two previous lines are spaces. So, the first two presses remove auto-indent and the third press executes the program.

Auto-completion

REPL provides an auto-completion feature that enables you to write your programs faster. If the line that you have typed resembles the beginning of the name of something, you can press *tab* so that REPL can display possible things that you can enter. For instance, if your type m and press the *tab* button, the letter expands to machine. If you enter a dot(.) after machine and press *tab*, you should receive several options, as shown here:

```
>>> machine.
__name__        info            unique_id       reset
bootloader      freq            rng             idle
sleep           deepsleep       disable_irq     enable_irq
Pin
```

You can expand the word machine as much as possible. For instance, type `macine.pin.AF3` and press *tab*. This will expand to `macine.pin.AF3_TIM`. If you press *tab* again, REPL will provide you with a list of possible expansions, as shown in the following code:

```
>>> machine.Pin.AF3_TIM
AF3_TIM10        AF3_TIM11        AF3_TIM8        AF3_TIM9
>>> machine.Pin.AF3_TIM
```

Interrupting a running program

A running program can be interrupted by simply pressing *Ctrl+C*. This executes a keyboard interrupt which takes you back to REPL, provided the program does not intercept the keyboard interrupt exception. Check the following example:

```
>>> for i in range(1000000):
...     print(i)
...
0
1
2
3
...
6466
6467
6468
Traceback (most recent call last):
  File "<stdin>", line 2, in <module>
KeyboardInterrupt:
>>>
```

Paste mode

Pasting code directly into the terminal window messes up the auto-indent function, as shown in the following example.

If you copy this piece of code:

```
def foo():
    print('This is a test to show paste mode')
    print('Here is a second line')
foo()
```

And try to paste it into REPL, it will appear as shown here:

```
>>> def foo():
...         print('This is a test to show paste mode')
...             print('Here is a second line')
...             foo()
...
  File "<stdin>", line 3
IndentationError: unexpected indent
```

The correct way to do it is to press *Ctrl+E* to enter `paste mode`. This turns off auto-indent and replaces the >>> prompt with ===.

```
>>>
paste mode; Ctrl-C to cancel, Ctrl-D to finish
=== def foo():
===     print('This is a test to show paste mode')
===     print('Here is a second line')
=== foo()
===
This is a test to show paste mode
Here is a second line
>>>
```

To exit paste mode, you press *Ctrl+D*. The program gets compiled immediately after exiting paste mode.

Soft reset

A soft reset resets the Python interpreter without affecting the connection between the MicroPython board and REPL. A soft reset can be done by pressing *Ctrl+D* or by executing the following line in the Python code:

```
raise SystemExit
```

For instance, if you reset your MicroPython board and run a `dir()` command, you will see something same as this:

```
>>> dir()
['__name__', 'pyb']
```

Go ahead and create some variables and run the `dir()` command once more. REPL will appear as shown here:

```
>>> i = 1
>>> j = 23
>>> x = 'abc'
>>> dir()
['j', 'x', '__name__', 'pyb', 'i']
>>>
```

If you press *Ctrl+D* and repeat the `dir()` command, you will notice that the variables you created are no longer there:

```
PYB: sync filesystems
PYB: soft reboot
MicroPython v1.5-51-g6f70283-dirty on 2015-10-30; PYBv1.0 with STM32F405RG
Type "help()" for more information.
>>> dir()
['__name__', 'pyb']
>>>
```

The special variable _ (underscore)

When you perform computations using REPL, the results of the previous statement are stored in the _ (underscore) variable, as shown in the following code:

```
>>> 1 + 2 + 3 + 4 + 5
15
>>> x = _
>>> x
15
>>>
```

Raw mode

You can enter raw mode by pressing *Ctrl+A*. It is not widely used and is usually intended for programmatic use. Once you enter raw mode, you can send code followed by *Ctrl+D*. *Ctrl+D* is acknowledged by **OK**, after which the code is compiled and executed. To leave raw mode, press *Ctrl+B*.

See also

If you feel ready to tackle your first MicroPython project, go to the next recipe. You will learn how to configure your ESP8266 so that you can start programming it using MicroPython.

Getting started with MicroPython on the ESP8266

There are several things you have to set up before using MicroPython to program your ESP8266 board. We will be going through the setup process in this recipe. This way you will you know how to configure the ESP8266 board to be used with MicroPython.

Getting ready

All you need in this stage is your ESP8266 and a USB cable. Connect your ESP8266 board to your computer.

How to do it...

1. The first thing you should do is download the latest MicroPython firmware from the following link: `http://micropython.org/download`.

2. The .bin files for the MicroPython port for ESP8266 are under the **Firmware for ESP8266 boards** subheading. Download the latest stable firmware build.

3. The next step is deploying the firmware. To do this, the ESP8266 chip has to be put in bootloader mode and then you copy across the firmware. The procedure for doing that depends on the board you are using. However, for the Adafruit Feather HUZZAH and NodeMCU boards, which have a USB connector and properly connected DTR and RTS pins, the process is easy as all those steps are done automatically.

> To get the best results, it is advisable to erase the entire flash for your ESP8266 board before uploading new MicroPython firmware. `esptool.py` is used to copy across the firmware. Install it using `pip` (version 1.2.1 or later) with this command:
>
> `pip install esptool`

4. Then use `esptool.py` to erase the flash:

   `esptool.py --port /dev/ttyUSB0 erase_flash`

5. You may have to change the serial port in your command to the serial port that your ESP8266 board is connected to. If you do not know the serial port number of your ESP8266, you can check in the Arduino IDE. Just open the IDE and then click on **Tools | Ports**. You should see the serial port of your ESP8266 board listed there. Replace the serial port in the command (`/dev/ttyUSB0`) with the serial port of your board.

6. Once that is sorted out, install the firmware with this command:

```
esptool.py --port /dev/ttyUSB0 --baud 460800 write_flash --flash_
size=detect 0 esp8266-2016-05-03-v1.8.bin
```

7. Make sure you change the name of the firmware `.bin` file in the command (`esp8266-2016-05-03-v1.8`) to that of the firmware you downloaded.

8. Once the firmware is successfully installed on your ESP8266 board, you can access REPL on your board via a wired connection (UART serial port) or through Wi-Fi.

Accessing REPL through the serial port is a common practice. REPL is usually available on the UART0 serial port, which is on GPIO1 (TX) and GPIO3 (RX). The default baud rate is 115,200. If the board you are using has a USB to serial converter, you should be able to access REPL directly from your computer, otherwise you have to use an external serial to USB converter or any other way that you are comfortable with.

Accessing REPL through USB serial is done using a terminal emulator, such as Windows Tera Term, picocom or minicom on Linux, or the built-in screen program on a mac. There are many other serial terminal programs that you can still use. The choice is yours.

Here is an example of how you connect to the ESP8266 serial port on Linux:

```
picocom /dev/ttyUSB0
```

When you access REPL via a serial terminal, you can test if the serial port is working by hitting enter a few times. If the serial port connection is good, the REPL prompt (>>>) should be displayed on the screen.

How it works...

MicroPython support for the ESP8266 chip is excellent. MicroPython allows you to access all functions and interfaces of the ESP8266 such as GPIO pins, PWM, ADC, I2C, and SPI. In addition to this, MicroPython supports Wi-Fi and Internet access, and also comes with a web-based REPL that enables you to run code on the ESP8266 through your web browser.

To use MicroPython, you have to connect to the ESP8266 serial port using a terminal window at a baud rate of 115200. Most serial terminals will work. Once the serial terminal is connected, you can enter your MicroPython code in the REPL.

It is important to remember that MicroPython is not exactly the same as the Python you use on your desktop (CPython). Most of the CPython standard library is not available. Therefore, you'll need to check the MicroPython documentation so that you can find out which functions are supported.

If you want to learn how to control, GPIO pins of the ESP8266 using MicroPython, proceed to the next recipe.

Controlling pins using MicroPython

In this recipe, we will learn how to control the ESP8266 pins with MicroPython. To do that we will come up with a setup where we will be switching the state of an LED connected to an ESP8266 board GPIO pin. This will help you understand how to control digital outputs using MicroPython.

Getting ready

You will need the following things to accomplish the setup:

- ESP8266 board (Adafruit HUZZAH)
- USB cable
- LED (https://www.sparkfun.com/products/528)
- 220 Ω resistor (https://www.sparkfun.com/products/10969)

Begin by mounting the LED onto the breadboard. Connect one end of the 220 Ω resistor to the positive leg of the LED (the positive leg of an LED is usually the taller one of the two legs). Connect the other end of the resistor to pin 5 of the ESP8266 board. Then connect the negative leg of the LED to the **GND** pin of the ESP8266 board. The connection is as shown in the following diagram:

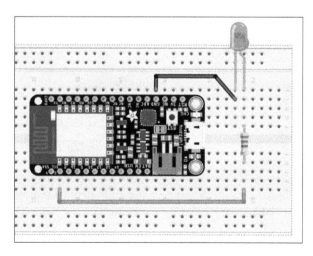

Once the setup is complete, connect the ESP8266 board to your computer via a USB cable.

How to do it...

1. Using terminal software, connect to the ESP8266 board via serial.

2. Then proceed to type the following command:

    ```
    import machine
    ```

3. Then type:

    ```
    pin = machine.Pin(5, machine.Pin.OUT)
    ```

4. After that, type:

    ```
    pin.value(1)
    ```

5. The LED should light up.

How it works...

You can control other components with your ESP8266 through its GPIO pins. Not all the pins are available for use though. In many boards, only pins 0, 2, 4, 5, 12, 13, 14, 15, and 16 are usually available.

The machine module holds the pin configurations and modes. Therefore, it is imported first before proceeding to the next code. That is done in this line:

```
import machine
```

The next step is configuring the pin that we want to use. We define the pin number and the mode we want it to work in. It can be an input or an output. Inputs are denoted by IN by while the outputs are denoted by OUT.

You can also determine whether the pin is connected to a pull-up resistor or not. If you want to configure the pin as connected to a pull-up resistor, you use the syntax PULL_UP. If the pin is not connected to the internal pull-up resistor, you do not have to specify that configuration. That way the pin defaults to None, which means that it is not connected to a pull-up resistor. An example of code that configures pin 0 as an input connected to the internal pull-up resistor is shown here:

```
>>> pin = machine.Pin(0, machine.Pin.IN, machine.Pin.PULL_UP)
```

In our case, we are using pin 5 as an output with the internal pull-up resistor not connected. This is set by this line of code:

```
pin = machine.Pin(5, machine.Pin.OUT)
```

The final step is setting the state of output pin 5. This is accomplished using the `value()` function. The value of the pin can either be set to either 0 or 1. 0 represents an off state while 1 represents an on state. In our code, this is accomplished by the line:

```
pin.value(1)
```

This turns the LED connected to pin 5 on. This command can also be implemented using the `pin.high()` statement to turn on the LED. `pin.low()` turns off the LED.

There is more...

Write code to turn off the LED.

See also

You can use MicroPython to read sensor data from an ESP8266. Proceed to the next recipe to learn how to do that.

Reading data from a sensor using MicroPython

Reading sensor data is an essential function of the ESP8266. You can do that using MicroPython. To demonstrate it, we are going to read the temperature and humidity from the DHT11 sensor using MicroPython.

Getting ready

In this tutorial, you will need an ESP8266 board, a USB cable, and a few other components, which include:

- DHT11 temperature/humidity sensor (https://www.sparkfun.com/products/10167)
- 10 kΩ resistor
- Breadboard
- Jumper wires

Mount the ESP8266 board and the DHT11 sensor onto the breadboard. Connect a 10 kΩ pull up resistor to the DHT11 data pin and connect the pin and **GND** pin to the 3V pin and GND pin of the ESP8266 board, respectively. Finally, connect the data pin of the DHT11 to GPIO 5 of the ESP8266 board. Use jumper wires to do the connections.

The setup is shown in the following diagram:

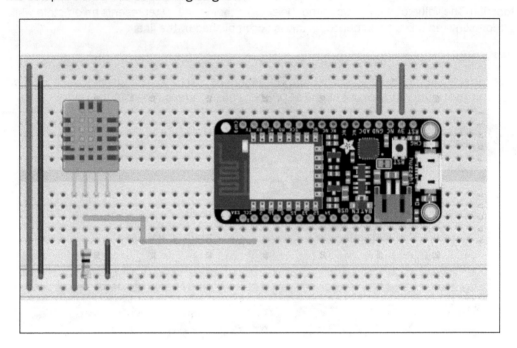

 Refer back to *Chapter 2, Your First ESP8266 Projects* to confirm the DHT11 pin configuration.

How to do it...

You can use MicroPython to read digital signals from the ESP8266 GPIO pins.

You do this by first importing the machine class, then configuring the GPIO pin as an input and printing its value on the serial terminal. Here is some example code:

```
import machine
p2 = machine.Pin(2, machine.Pin.IN)      # create input pin on GPIO2
print(p2.value())        # get value, 0 or 1
```

MicroPython can also be used to read analog signals from the analog pin. To do this, you first import the machine class, then construct an ADC pin object, then you read and print out the value of the pin with the read() function. The following example shows how to go about it:

```
import machine
adc = machine.ADC(0)
adc.read()
```

Though the DHT11 is a digital sensor, we cannot use the basic method of *Reading digital signals* as described previously. This is because the DHT11 sensor sends the digital readings to the ESP8266 using the 1-wire protocol. As such, we will need to import the DHT library so that it can handle the reading of data from the DHT11 sensor to obtain credible measurements. The code is as shown here.

The first part of the code constructs a DHT object with the digital pin that the DHT11 is connected to:

```
import dht
import machine
d = dht.DHT11(machine.Pin(5))
```

Then this is followed by reading the temperature and humidity values from the sensor:

```
d.measure()
d.temperature()
d.humidity()
```

How it works...

Digital Humidity and Temperature (**DHT**) sensors use thermistors and capacitive humidity sensors to measure the temperature and humidity of the surrounding air. They have a chip that performs the analog to digital conversion. The chip then outputs digital data via a 1-wire interface. There are some newer versions that use an I2C interface instead of the 1-wire interface.

The DHT11 sensor uses the 1-wire interface. Therefore, the custom 1-wire protocol is used to read the measurements. The read data consists of a temperature value, a humidity value, and a checksum. Since you are going to be using the 1-wire interface, the first line of your code should construct objects that refer to the data pin. This is implemented by the following lines of code:

```
import dht
import machine
d = dht.DHT11(machine.Pin(5))
```

Once the DHT object has been created, you can proceed to read the temperature and humidity data. The value returned by the temperature() function is in degrees Celsius. The value returned by the humidity() function is relative humidity given as a percentage. The following code implements that:

```
d.measure()
d.temperature()
d.humidity()
```

To get accurate results, read data from the DHT11 once every second. This gives the sensor time to process the measurements. Increasing the frequency of readings can lead to duplicate or corrupted values.

See also

Once you have gotten sensor data, you can transmit it to the cloud for remote monitoring. You will learn how to do this in the next recipe.

Sending data to the cloud using MicroPython

This recipe is going to introduce us to the IoT using MicroPython. We will use MicroPython to send measurement data to dweet.io from the ESP8266. Through that, you will learn some IoT basics and how to implement them using MicroPython.

Getting ready

The hardware setup will be the same as the one you used in the previous recipe. The only difference is that we will be connecting and sending data to dweet.io. dweet.io is a cloud server that you can use to easily publish and subscribe to data. It does not require you to sign up or set it up. All you need to do is publish and you are good to go.

A simple HAPI web API is used to send data from your thing to the cloud. Sending data is accomplished by calling a URL such as https://dweet.io/dweet/for/my-thing-name?hello=world.

Replace my-thing-name in the URL with the name of your choice and then proceed to log data online. The query parameters you add to the URL will be added as key value pairs to the dweet content.

How to do it...

1. To successfully log sensor data to dweet.io, you should first connect the ESP8266 board to a Wi-Fi network that has an Internet connection. Then read data from the DHT11 sensor and publish it to dweet.io. The following code will accomplish all these tasks.

   This function connects the ESP8266 to the Internet:

   ```
   # Function to connect to the WiFi
   def do_connect():
       import network
       sta_if = network.WLAN(network.STA_IF)
       if not sta_if.isconnected():
   ```

```
            print('connecting to network...')
            sta_if.active(True)
            sta_if.connect('<essid>', '<password>')
            while not sta_if.isconnected():
                pass
        print('network config:', sta_if.ifconfig())
```

2. Make sure you modify the `ssid` and `password` in the `sta_if.connect('<essid>', '<password>')` line of the code, so that they match those of the Wi-Fi network you want the ESP8266 to connect to.

   This function is for sending the `HTTP` request:

```python
# Function to send an HTTP request
def http_get(url):
    _, _, host, path = url.split('/', 3)
    addr = socket.getaddrinfo(host, 80)[0][-1]
    s = socket.socket()
    s.connect(addr)
    s.send(bytes('GET /%s HTTP/1.0\r\nHost: %s\r\n\r\n' %
    (path, host), 'utf8'))
    while True:
        data = s.recv(100)
        if data:
            print(str(data, 'utf8'), end='')
        else:
            break
```

3. Connect to the Internet by executing this function:

```python
# Connect to WiFi
do_connect();
```

4. Get temperature and humidity readings from the `DHT11` sensor:

```python
# Make measurements
import dht
import machine
d = dht.DHT11(machine.Pin(5))

d.measure();
temperature = d.temperature();
humidity = d.humidity();
```

5. Build a `http` request that will be sent to `dweet.io`:

```python
# Build request
url = 'https://dweet.io/dweet/for/myesp8266';
url += '?temperature=' + temperature;
url += '&humidity' + humidity;
```

6. Send the `http` request:

```
# Send request
http_get();
```

7. Set your prompt into paste mode by pressing *Ctrl+E* and then copy and paste the code to the prompt. You can get the complete code from this link: `https://github.com/marcoschwartz/esp8266-iot-cookbook`. Once you have successfully pasted the code, press *Ctrl+D* to exit paste mode and execute the code.

You should start seeing data being logged on your console. The data you see is the replies from `dweet.io` every time data is published.

How it works...

The code has two functions, which handle connecting to the Internet and publishing data to `dweet.io`. The function that connects the ESP8266 board to the Internet is called `do_connect()`. In this function, the network library is imported and a `network` object is created in STA mode. If the ESP8266 is not connected to a Wi-Fi network, the `network` object is set to active using the `active()` function and the board gets connected to the Wi-Fi network whose SSID and password have been provided. This is done by the `connect()` function. If the connection is successful, the ESP8266 prints out the network configuration on the prompt, otherwise nothing happens.

The function that publishes data to `dweet.io` is called `http_get(url)`. It starts by dividing the URL into its different parts, that is, host name and path, using the `split()` function. The program then uses the host name and socket to create an address, which the ESP8266 uses to connect to the server, using the `connect()` function. The `send()` function is then used to publish a `GET` request to the server. The ESP8266 then listens to the reply from the server, using the `recv()` function. If there is a reply, it is printed on the serial terminal.

The main part of the program begins with calling the `do_connect()` function so that the ESP8266 can connect to the Internet. Once the board successfully connects to the Internet, it takes the temperature and humidity readings from the DHT11 sensor and saves the readings in two variables: `temperature` and `humidity`. The `http` request is then built and the temperature and humidity readings are added to it. The `http_get()` function then publishes the temperature and humidity measurements to `dweet.io`.

See also

There are some problems you may run into when using MicroPython to program your ESP8266 board. Check out the next recipe for some of their causes and ways to solve them.

Troubleshooting common MicroPython issues

You may run into some of these issues when using MicroPython. Here is how to troubleshoot and deal with them.

A library can't be used

Sometimes you may find that a library cannot be used. If that happens, check to make sure the library has been installed correctly. Also make sure you have the latest version of the library.

The menu to upload files to the ESP8266 is not visible

If this occurs, make sure to follow all the instructions well. If the menu is still not visible after following the instructions, reboot your Arduino IDE.

The board can't be configured via OTA

If you cannot configure the board via OTA, follow all the instructions again. Also make sure that your board is connected to the right Wi-Fi network.

5
Cloud Data Monitoring

In this chapter, we will cover:

- ▶ Internet of Things platforms for the ESP8266
- ▶ Connecting sensors to your ESP8266 board
- ▶ Posting the sensor data online
- ▶ Retrieving your online data
- ▶ Securing your online data
- ▶ Monitoring sensor data from a cloud dashboard
- ▶ Creating automated alerts based on the measured data
- ▶ Monitoring several ESP8266 modules at once
- ▶ Troubleshooting common issues with web services

Introduction

In this chapter, we will be looking at cloud data monitoring using the ESP8266. The ESP8266 is an ideal chip for IoT projects since it offers Wi-Fi connectivity as well as basic GPIO pin functionality. To demonstrate all that, we will use the ESP8266 to read sensor data and log it to and retrieve it from online servers. We will also look at other different features of IoT platforms that we can use to secure and monitor our data.

Internet of Things platforms for the ESP8266

The ESP8266 module is a very popular chip in the IoT field. As such, there are many IoT platforms that are based on it. The platforms come with different features and configurations that facilitate the use of the ESP8266 in IoT projects. This recipe is going to look at some of the available platforms and their features and functionalities.

Sparkfun ESP8266 thing

The Sparkfun ESP8266 thing is a development board designed around the ESP8266 module. It comes with an integrated FTDI USB-to-serial chip, Li-Po battery charging circuit, Wi-Fi, 512 KB flash memory, power switch, and an LED. In addition to this, it has very good documentation that guides users on how to set it up and use it.

The Sparkfun thing can be used for basic GPIO pin functions, such as reading sensors and controlling outputs such as LEDs. It can also be used to post data to online servers. Sparkfun has an online platform where you can post your data. It is `http://data.sparkfun.com` and you can easily log data in to it using a Node.js-based data logging tool `http://phant.io`. The ESP8266 thing posts sensor data to `phant.io` for monitoring purposes. You can also log data to other online servers directly, though Sparkfun does not offer adequate support and documentation on that:

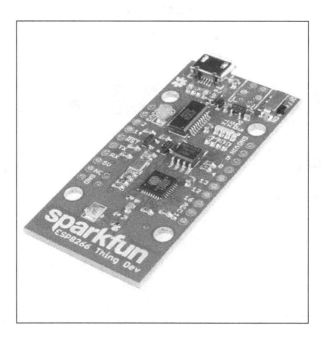

Adafruit feather HUZZAH

The Adafruit feather HUZZAH ESP8266 is one of the many feather family boards manufactured by Adafruit. Just like the Sparkfun thing, it is based on the ESP8266 module and comes with an integrated Li-Po battery charging circuit, an on-board LED, and a CP2104 USB-to-serial chip. It has a 4 MB flash memory, which is considerably more than the Sparkfun thing's 512 KB flash memory.

Although it is best programmed using the Arduino IDE, you can also program the HUZZAH ESP8266 with NodeMCU LUA or MicroPython. This is because there is ample flash memory to store the firmware builds of those other programming languages.

There are lots of materials and resources on the HUZZAH ESP8266 that one can use to learn how to program the board. Although there is no online platform offered by Adafruit for use with its IoT boards, the HUZZAH ESP8266 can be used with almost all the other online platforms available for IoT projects. A good example is dweet.io.

Since the HUZZAH ESP8266 is part of the feather boards family, you can easily interface it with sensors, actuators, and other boards in the feather family. This makes it more versatile and increases prototyping speed:

NodeMCU ESP8266

NodeMCU is an open source firmware and development kit. This board has a USB-to-serial chip, but lacks a Li-Po battery charging circuit and a Li-Po battery connector. It comes with a NodeMCU LUA firmware build that allows it to be programmed using the LUA programming language, alt though it can still be programmed using the Arduino IDE and PlatformIO.

The NodeMCU development kit comes with 4 MB of flash memory, so it can support MicroPython, in addition to LUA and the Arduino core. However, due to the lack of a Li-Po battery connector and charging circuit, using the NodeMCU devkit away from a dedicated power supply such as a PC is not always easy. Moreover, there are not many sensors, actuators, and boards that have been designed to match its pin configuration. This makes prototyping a little bit difficult:

Wio Link

The Wio Link event kit is an event-monitoring kit based on the ESP8266 chip. The platform has a mobile phone application called Wio Link from which you can configure the different connected sensors/actuators and read data from them. With Wio Link, you do not need to program your board with an Arduino IDE, as the app can program it via OTA updates. The Wio Link board is compatible with Seed Studio's Groove line of products. This makes it easy to connect hardware components to it.

Just like the HUZZAH ESP8266, the Wio Link board has a Li-Po battery connector and charging circuit, which makes it ideal for portable projects. It also comes with an on-board LED that enables you to see the state of the board:

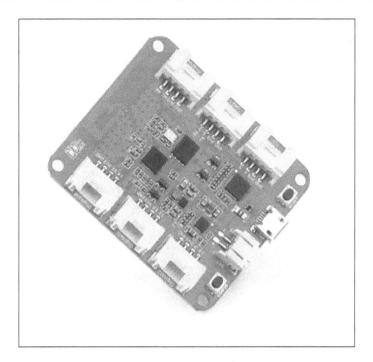

See also

Having understood the different IoT platforms that are available for the ESP8266, you can now start working on your first IoT project. The next recipe will start you off by guiding you through how to connect sensors to your ESP8266 board.

Connecting sensors to your ESP8266 board

In this recipe, we are going to look at how to connect sensors to an ESP8266 board. There are two kinds of sensors that you will be using with your ESP8266: digital sensors and analog sensors.

The digital sensors output digital signals that you can read using the ESP8266 digital GPIO pins. The digital sensor outputs have only two states: high (logic 1) and low (logic 0). The high signals are ideally at a voltage level of 3.3V, while the low signals are at a voltage level of 0V.

Analog sensors output analog signals. The output comes in various voltage levels between 0V and 3.3V. The output signal is read using the ESP8266 analog pin (labeled ADC on our board).

It is advisable not to connect the analog sensor output directly to the ESP8266 analog pin. This is because the sensor output voltage range of 0V-3.3V is greater than the input voltage range of the analog pin, which is 0V-1V. The best way to connect an analog sensor to the analog pin is via a voltage divider. The voltage divider steps down the sensor output voltage range to the desired 0V-1V.

On the ESP8266, analog readings from the sensor vary between 0 and 1023. This is because the ESP8266 has a 10-bit analog-to-digital converter. A 0 reading corresponds to 0V on the analog input pin, while a 1023 reading corresponds to 1V on the analog input pin.

To demonstrate how to use sensors on your board, we will connect a DHT11 sensor and a soil moisture sensor the Adafruit HUZZAH feather board. The DHT11 is a digital sensor, while the soil moisture sensor is an analog sensor.

Getting ready

You will need the following components to complete the setup:

- ESP8266 board
- USB cable
- DHT11 (`https://www.sparkfun.com/products/10167`)
- Soil moisture sensor (`https://www.sparkfun.com/products/13322`)
- 220 Ω resistor
- 100 Ω resistor
- 10 kΩ resistor
- Breadboard
- Jumper wires

1. Start by mounting the ESP8266 board and the DHT11 sensor onto the breadboard.
2. Then connect a 10 kΩ pull up resistor to the DHT11 data pin and connect the **VCC** pin and **GND** pin of the sensor to the **3V** pin and **GND** pin of the ESP8266 board, respectively.
3. Finally, connect the data pin of the DHT11 to GPIO 2 of the ESP8266 board.

   Now you can proceed to set up the soil moisture sensor:

4. Begin by connecting the soil moisture sensor **VCC** and **GND** pins to the ESP8266 board **3V** and **GND** pins respectively.
5. Then connect the **signal (SIG)** pin to the voltage divider. The voltage divider will be constructed using the **220** Ω and **100** Ω resistors.

6. Connect the output of the voltage divider to the analog pin. The voltage divider schematic diagram is shown here:

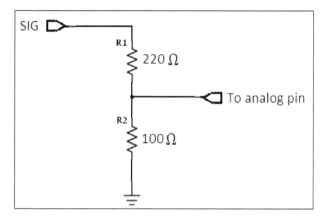

The complete setup will look like this:

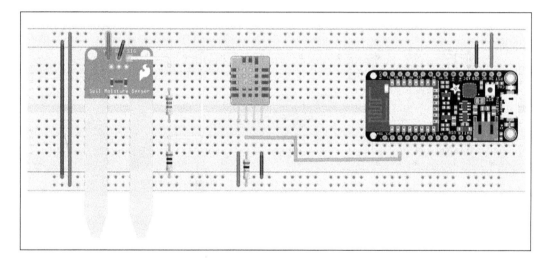

How to do it...

You will first get readings from the DHT11 sensor, then proceed to get readings from the soil moisture sensor. The soil moisture sensor reading is obtained from the analog pin using the analogRead() function.

Digital input is normally read using the `digitalRead()` function, but in the case of the `DHT11` sensor, some more data processing is required in order to get comprehensive readings. This processing is done by functions in the `DHT` library. You will use the `readTemperature()` and `readHumidity()` functions of the `DHT` library to obtain the temperature and humidity readings from the sensor. All the readings will be displayed on the serial monitor:

```
#include "DHT.h"

#define DHTPIN 2      // what digital pin we're connected to
#define DHTTYPE DHT11    // DHT 11

DHT dht(DHTPIN, DHTTYPE);

int sensorReading = 0; // holds value soil moisture sensor reading

void setup() {
  Serial.begin(9600);
  dht.begin();
}

void loop() {
  // Wait a few seconds between measurements.
  delay(2000);
  // get humidity reading
  float h = dht.readHumidity();
  // get temperature reading in Celsius
  float t = dht.readTemperature();
  // Check if any reads failed and exit early (to try again).
  if (isnan(h) || isnan(t)) {
    Serial.println("Failed to read from DHT sensor!");
    return;
  }
  //get soil moisture reading
  sensorReading = analogRead(A0);

  // display data on serial monitor
  Serial.print("Humidity: ");
  Serial.print(h);
  Serial.print(" %\t");
  Serial.print("Temperature: ");
  Serial.print(t);
  Serial.println(" *C ");
  Serial.print("Soil moisture: ");
  Serial.println(sensorReading);
}
```

1. Copy the sketch and paste it in your Arduino IDE.

2. Check to ensure that the ESP8266 board is connected. Select the board that you are using in the **Tools | Board menu** (in this case it is the Adafruit HUZZAH ESP8266).

3. Select the serial port your board is connected to from the **Tools | Port menu** and then upload the code. You can open the serial monitor to view the sensor data.

How it works...

The program starts by including the DHT library. The digital pin that the signal pin of the sensor is connected to and type of DHT sensor that is being used are then defined. A DHT object is created and the variable for holding the soil moisture sensor, sensorReading, is declared. The serial interface and DHT object are then initialized in the setup section of the sketch.

The loop segment handles reading of the sensor input. It starts with a 2-second delay that gives the sensor time to get readings. The humidity is then read using the dht.readHumidity() function and the temperature readings are acquired using the dht.readTemperature() function, which returns readings in degrees Celsius.

The sketch sends an alert if any of the DHT readings are not obtained and tries again until it receives valid readings. If the obtained readings are alright, the soil moisture sensor input is read. Then all the readings obtained from the sensors are displayed on the serial monitor.

There's more...

Use the digitalRead() function to read digital sensor input from the digital pins of the ESP8266. You can use a push button as the sensor.

See also

Wouldn't it be cool if you could post your sensor readings online? Well, the good news is that it is possible and that you can learn how to do it in the next recipe.

Posting the sensor data online

Using the ESP8266, you can log sensor data to online servers for monitoring and control purposes. In this recipe, we will be looking at how to do that. We will use an ESP8266 board to post and store sensor data on dweet.io.

Getting ready

To do this tutorial, you will need an ESP8266 board and a USB cable among several other things, which include:

- DHT11 (`https://www.sparkfun.com/products/10167`)
- Soil moisture sensor (`https://www.sparkfun.com/products/13322`)
- 220 Ω resistor
- 100 Ω resistor
- 10 kΩ resistor
- Breadboard
- Jumper wires

The hardware setup will resemble the one in the previous recipe.

In addition to setting up the hardware, you will set up the online platform where the sensor data will be posted. Luckily the platform we are using, `dweet.io`, is simple to use and requires no setup or signup. All you need to do is choose a name for your thing, which in this case is your ESP8266, then you can start publishing data to `dweet.io`.

How to do it...

Posting data to `dweet.io` is simple. It is done by calling a URL such as `https://dweet.io/dweet/for/my-thing-name?hello=world`. You are required to change `my-thing-name` to the name of your thing, `hello` to the name of the parameter that you are posting, and `world` to the value of the parameter that you are posting online.

In our case, we will call our thing `garden-monitor-11447`, and the names of the parameters we will be posting are `humidity`, `temperature`, and `moisture`. Our URL will look like this: `https://dweet.io/dweet/for/garden-monitor-11447?humidity=<humidity value>&temperature=<temperature value>&moisture=<moisture value>`.

Once you have successfully generated the URL:

1. Connect the ESP8266 board to a Wi-Fi network that has Internet connection and then read the `humidity`, `temperature`, and `soil moisture` values from the sensors, as done in the previous recipe.
2. Replace the `<humidity value>`, `<temperature value>`, and `<moisture value>` in the preceding URL with the readings obtained from the sensors. The URL will be used to send a `http` request to the `dweet.io` server.

3. When the ESP8266 successfully sends the `http` request, the sensor data will be published on the `dweet.io` platform. To view the data, visit this URL from any web browser: `https://dweet.io/follow/garden-monitor-11447`:

```
// Libraries
#include <ESP8266WiFi.h>
#include "DHT.h"
```

4. `ssid` and `password` is set in this section:

```
// Wi-Fi network SSID and password
const char* ssid     = "your-ssid";
const char* password = "your-password";
```

5. Store the `host` name of the cloud server:

```
// Host
const char* host = "dweet.io";
```

6. Define the pin that is connected to the `DHT11` signal pin and the type of `DHT` sensor that we are using:

```
#define DHTPIN 2 // what digital pin DHT11 is connected to
#define DHTTYPE DHT11 // DHT 11 sensor
```

7. Create a `dht` object:

```
DHT dht(DHTPIN, DHTTYPE);
```

8. Declare a variable that will hold the `moistureReading`:

```
int moistureReading = 0; // holds value soil moisture sensor
reading
```

9. Initialize the serial interface and the `DHT` object. Configure the `ssid` and `password` of the Wi-Fi network and connect the ESP8266 to it:

```
void setup() {
  Serial.begin(115200); // initialize serial interface
  dht.begin(); // initialize DHT11 sensor
  delay(10);

  // We start by connecting to a WiFi network
  Serial.println();
  Serial.println();
  Serial.print("Connecting to ");
  Serial.println(ssid);

  WiFi.begin(ssid, password);

  while (WiFi.status() != WL_CONNECTED) {
```

```
        delay(500);
        Serial.print(".");
    }


    Serial.println("");
    Serial.println("WiFi connected");
    Serial.println("IP address: ");
    Serial.println(WiFi.localIP());
}
```

10. Delay for five seconds, then print the name of the host we are connecting to on the serial monitor:

```
void loop() {
  delay(5000);

  Serial.print("connecting to ");
  Serial.println(host);
```

11. Connect to the `host` server. Retry if not successful:

```
// Use WiFiClient class to create TCP connections
WiFiClient client;
const int httpPort = 80;
if (!client.connect(host, httpPort)) {
  Serial.println("connection failed");
  return;
}
```

12. Get readings from the `DHT11` sensor and `soil moisture` sensor:

```
// Read sensor inputs
// get humidity reading
float h = dht.readHumidity();
// get temperature reading in Celsius
float t = dht.readTemperature();
// Check if any reads failed and exit early (to try again).
while (isnan(h) || isnan(t)) {
  Serial.println("Failed to read from DHT sensor!");
  delay(2000); // delay before next measurements
  //get the measurements once more
  h = dht.readHumidity();
  t = dht.readTemperature();
}
//get soil moisture reading
moistureReading = analogRead(A0);
```

- ❏ Generate a URL for the GET request we will send to the host server. The URL will include the sensor readings:

```
// We now create a URI for the request
String url = "/dweet/for/garden-monitor-11447?humidity=";
url += String(h);
url += "&temperature=";
url += String(t);
url += "&moisture=";
url += String(moistureReading);
```

13. Send the GET request to the server and check whether the request has been received, or if it has been timed out:

```
// Send request
Serial.print("Requesting URL: ");
Serial.println(url);

client.print(String("GET ") + url + " HTTP/1.1\r\n" +
             "Host: " + host + "\r\n" +
             "Connection: close\r\n\r\n");
unsigned long timeout = millis();
while (client.available() == 0) {
  if (millis() - timeout > 5000) {
    Serial.println(">>> Client Timeout !");
    client.stop();
    return;
  }
}
```

- ❏ Read incoming data from the host server line by line and display the data on the serial monitor. Close the connection after all the data has been received from the server:

```
// Read all the lines from the answer
while(client.available()){
  String line = client.readStringUntil('\r');
  Serial.print(line);
}

// Close connecting
Serial.println();
Serial.println("closing connection");
}
```

14. Copy the sketch to your Arduino IDE and change the `ssid` in the code from `your-ssid` to the name of your Wi-Fi network, and the `password` from `your-password` to the password of your Wi-Fi network.

15. Upload the sketch to your ESP8266 board. Open the serial monitor so that you can view the incoming data.

How it works...

The program connects to the Wi-Fi network using the provided `password` and `ssid`. It then proceeds to connect to the provided cloud/host server using the `client.connect()` function. Once the ESP8266 connects successfully, data from the sensors is read and a URL is generated that includes the updated sensor data. The URL is then sent to the host server using the `client.print()` function.

Once the data has been successfully sent, the sketch waits for a reply from the server. It does this with the `client.available()` function, which checks whether there is incoming data from the server. If there is data available, the sketch reads it and displays it on the serial monitor. The ESP8266 posts sensor data to `dweet.io` every five seconds.

See also

Once the sensor data is posted online, you will need to retrieve it for monitoring and control purposes. This can be done using the ESP8266, as explained in the next recipe.

Retrieving your online data

The ESP8266 can be used to retrieve data from online servers so long as it is connected to the Internet. To demonstrate how to do that, we will use our ESP8266 module to retrieve the most recent sensor data that was posted to `dweet.io`, in our previous recipe. This will give you an idea of how to go about retrieving data from online sources.

Getting ready

For this recipe, you will only need your ESP8266 board and a USB cable. You can also leave your setup as it was in the second recipe in this chapter, although we won't need to read sensor input this time round.

How to do it...

Reading the most recent data from `dweet.io` is just as simple as posting it. All you've got to do is send an `http` request using the following URL: `https://dweet.io/get/latest/dweet/for/my-thing-name`:

1. Replace `my-thing-name` with the name of your thing and you are good to go. Therefore, in our case, the URL we will use is `https://dweet.io/get/latest/dweet/for/garden-monitor-11447`.

2. In addition to getting the most recently posted data, you can retrieve the last five data entries in a 24-hour period. You can do that using the following URL: `https://dweet.io/get/dweets/for/garden-monitor-11447`. However, we will not be using this URL in our code.

3. Now that you have your URL figured out, connect your ESP8266 board to a Wi-Fi network with an Internet connection and send an `http` request to `dweet.io` using your URL. `dweet.io` will send a response that contains the most recently logged sensor data.

4. You can capture and display the data on the serial monitor or use it to perform a certain function on your board. The code we are using will display the data on the serial monitor:

    ```
    // Libraries
    #include <ESP8266WiFi.h>
    ```

5. `ssid` and `password`:

    ```
    // SSID
    const char* ssid     = "your-ssid";
    const char* password = "your-password";
    ```

6. Store the `host` name of the cloud server:

    ```
    // Host
    const char* host = "dweet.io";
    ```

7. Configure the `ssid` and `password` of the Wi-Fi network and connect the ESP8266 to the Wi-Fi network:

    ```
    void setup() {

      // Serial
      Serial.begin(115200);
      delay(10);

      // We start by connecting to a WiFi network
      Serial.println();
      Serial.println();
      Serial.print("Connecting to ");
    ```

```
     Serial.println(ssid);

     WiFi.begin(ssid, password);

     while (WiFi.status() != WL_CONNECTED) {
       delay(500);
       Serial.print(".");
     }

     Serial.println("");
     Serial.println("WiFi connected");
     Serial.println("IP address: ");
     Serial.println(WiFi.localIP());
   }
```

8. Delay for five seconds then print the name of the host we are connecting to on the serial monitor:

```
   void loop() {

     delay(5000);

     Serial.print("connecting to ");
     Serial.println(host);
```

9. Connect to the `host` server:

```
     // Use WiFiClient class to create TCP connections
     WiFiClient client;
     const int httpPort = 80;
     if (!client.connect(host, httpPort)) {
       Serial.println("connection failed");
       return;
     }
```

10. Generate the `URI` for the `GET` request that we will send to the `host` server. We will use it to retrieve data from the `dweet.io` servers:

```
     // We now create a URI for the request
     String url = "/get/latest/dweet/for/garden-monitor-11447";
```

11. Send the `GET` request to the server and check whether the request has been received, or if it has been timed out:

```
     // Send request
     Serial.print("Requesting URL: ");
     Serial.println(url);
```

```
client.print(String("GET ") + url + " HTTP/1.1\r\n" +
             "Host: " + host + "\r\n" +
             "Connection: close\r\n\r\n");
unsigned long timeout = millis();
while (client.available() == 0) {
  if (millis() - timeout > 5000) {
    Serial.println(">>> Client Timeout !");
    client.stop();
    return;
  }
}
```

❑ Read incoming data from the host server line by line and display the data on the serial monitor. Close the connection after all the data has been received from the server:

```
// Read all the lines from the answer
while(client.available()){
  String line = client.readStringUntil('\r');
  Serial.print(line);
}

// Close connecting
Serial.println();
Serial.println("closing connection");
}
```

12. Copy the sketch to your Arduino IDE and change the `ssid` in the code from `your-ssid` to the name of your Wi-Fi network, and change the `password` from `your-password` to the password of your Wi-Fi network.

13. Upload the sketch to your ESP8266 board and open the serial monitor so that you can view the incoming data.

How it works...

The program connects to the Wi-Fi network using the provided `password` and `ssid`. It then proceeds to connect to the provided cloud/host server using the `client.connect()` function and sends the provided `URI` to the `host` server using the `client.print()` function.

Once the data has been successfully sent, the sketch waits for a reply from the server. It does this with the `client.available()` function, which checks whether there is incoming data from the server. If there is data available, the sketch reads it and displays it on the serial monitor.

The incoming data will be in the following format:

```
{
  "this": "succeeded",
  "by": "dweeting",
  "the": "dweet",
  "with": {
    "thing": "<my-thing-name>",
    "created": "<date of creation>",
    "content": {
      "Parameter1": "<value>",
      "Parameter2": "<value>",


      "ParameterN": "<value>"
    }
  }
}
```

 Everything enclosed in these brackets < > will be replaced by the actual values that are determined by the way you saved your data.

In our case, we should receive something like this:

```
{
  "this": "succeeded",
  "by": "dweeting",
  "the": "dweet",
  "with": {
    "thing": "garden-monitor-11447",
    "created": " 2017-01-18T15:36:59.791Z",
    "content": {
      "humidity": "35",
      "temperature": "22",
      "moisture": "512"
    }
  }
}
```

However, in reality `dweet.io` does not format the data as represented earlier. So, do not worry if you get something like this:

```
{"this": "succeeded", "by": "dweeting", "the": "dweet", "with":
{"thing": "garden-monitor-11447","created": " 2017-01-18T15:36:59.791
Z","content": {"humidity": 35",        "temperature": "22","moisture":
"512"}}}
```

It is the same data, without formatting.

If you want to use the sensor data values to do something on your board, use some string manipulation functions to strip away the excess characters and keep the sensor data. You will need an elaborate algorithm to do that.

There's more...

Use `https://dweet.io/get/dweets/for/garden-monitor-11447` to get the five most recently posted sensor readings.

See also

The way `dweet.io` is designed, it allows anyone to access your data so long as he or she knows the name of your thing. This may raise security concerns when dealing with sensitive data. To curb this, `dweet.io` offers a feature that allows you to keep your data secure. Our next recipe will show you how to use that feature.

Securing your online data

`dweet.io` secures your data through a method known as locking. Locks keep your data secure by preventing access, unless a thing or a person provides a special key. In this recipe, we will look at how to set up a lock and some features that come with it. This way you will know how to secure the data you post online from your ESP8266.

Locking your things

When you lock your thing, its name cannot be used by another person. You can acquire and maintain a lock by paying a small fee of $1.99 every month. Once you pay, you are provided with a lock and a special key. Using the lock and key, you can secure your things and prevent other people/things from accessing your data on `dweet.io`.

When you pay for a lock, you receive your lock ID and a unique key in your e-mail account. You can then lock your things through the web API by calling this URL:

`https://dweet.io/lock/{thing_name}?lock={your_lock}&key={your_key}`

If your thing is locked successfully, you will receive this reply:

```
{"this":"succeeded","by":"locking","the":"garden-monitor-
11447","with":"lock"}
```

You can also use the `https://dweet.io/locks` webpage to lock your things. All you need to do is provide the name of your thing, your lock, and the special key. The form looks like this:

Once your thing is locked, you will only be able to access it using the unique key. To do that, just add a parameter called key to the API call you are making. For instance:

```
https://dweet.io/dweet/for/{my_locked_thing}?key={my_
key}&hello=world&foo=bar
```

```
https://dweet.io/get/dweets/for/{my_locked_thing}?key={my_key}
```

```
https://dweet.io/get/latest/dweet/for/{my_locked_thing}?key={my_key}
```

```
https://dweet.io/listen/for/dweets/from/{my_locked_thing}?key={my_
key}
```

You can unlock your thing using this URL: `https://dweet.io/unlock/{thing_name}?key={your_key}`, or remove the lock completely regardless of what things it is connected to using this URL: `https://dweet.io/remove/lock/{your_lock}?key={your_key}`

In addition to data security, there are several other benefits that come with acquiring a lock. For instance, you get a 30-day storage period for every locked dweet, as opposed to the 24-hour storage period that is available for unlocked dweets. You can also access all data that was posted within a specific hour of any day in the 30-day period. This allows you more control over how you are going to monitor your processes. You can do that using this URL: `https://dweet.io/get/stored/dweets/for/{thing}?key={key}&date={date}&hour={hour}&responseType={type}`,

Where:

- ▶ {thing}: This is a valid name for your thing.
- ▶ {key}: This is the valid key for your locked dweet.
- ▶ {date}: This is the calendar date (YYYY-MM-DD) of the day you want to query data from.
- ▶ {hour}: This is the hour of the day that you want to start querying the data from. It is in the 24-hour format. Data that was published within one hour of the specified time will be returned.
- ▶ {type}: This can either be json or csv. If left it blank, the type will, by default be json. csv only works with dweets that have a static data model.

An example of a URL in that format is:

```
https://dweet.io/get/stored/dweets/for/my_thing?key=abc123&date=2016-
01-31&hour=13&responseType=csv
```

You can add alerts to locked dweets. The alerts let you know when a certain condition has been met by the data that is being posted on dweet.io. To set alerts, you use the following URL:

```
https://dweet.io/alert/{recipients}/when/{thing}/
{condition}?key={key}
```

You can remove alerts using this URL format:

```
https://dweet.io/remove/alert/for/{thing}?key={key}
```

Another feature that is not specifically accessible to locked dweets is real-time streams. By making a call to this URL: `https://dweet.io/listen/for/dweets/from/{my-thing-name}`, you will be able to receive dweets as they arrive.

There are different ways of ensuring that data stored on online servers is secure. The methods available to you will be determined by the online IoT platform you use. Therefore, find out whether your platform provides data security before you start using it.

See also

To employ better data presentation for monitoring trends, you need graphs and pie charts. You won't be able to access any of that on your ESP8266 board, but you can use cloud dashboards to do it. Our next recipe will show you how to go about it.

There are also other cloud storage services you can check out that would work with the ESP8266, such as ThingSpeak.

Monitoring sensor data from a cloud dashboard

In this recipe, we will be looking at how to use a graphical interface to monitor the data logged by the ESP8266. One of the main reasons why sensor data is posted online is so that you can easily monitor the data over time. This is facilitated by cloud dashboards that take the posted data and convert it to meaningful graphical presentations. The graphical presentations guide us in the decisions and actions we take pertaining to the system that is being monitored. We will use `dweet.io` and `freeboard.io` to demonstrate how cloud dashboards work.

Getting ready

Set up the hardware the same way you did in the, *Connecting sensors to your ESP8266 board* ercipe and upload the code for posting sensor data to `dweet.io`.

 That is the code we used in the, *Posting the sensor data online recipe.*

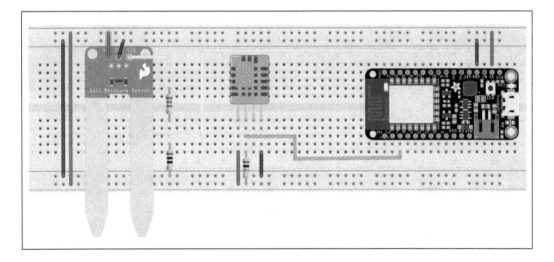

No configuration will be required in `dweet.io` for you to get a graphical view of the posted data. On the other hand, you will need to set up a few things in `freeboard.io` before you can start using it.

Start by creating an account on `freeboard.io`. You can try out the free version, since it has all the features we need for now:

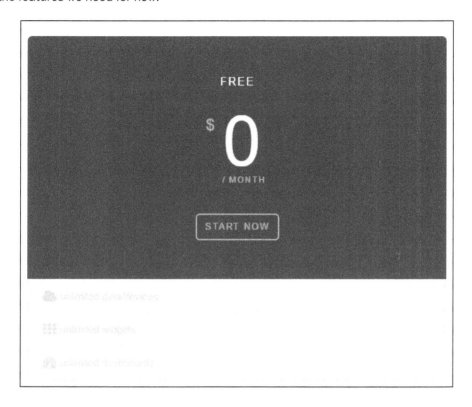

Click on **START NOW** to begin the signup procedure. You will be required to provide your preferred user name, your e-mail address, and a password, then you can complete the signup process. Now `freeboard.io` is ready to use.

How to do it...

`dweet.io` has a very easy-to-use dashboard.

All you need to do is call `http://dweet.io/follow/{thing-name}`. Replace `{thing-name}` with the name of your thing. It should open a window and plot the sent data in line graphs, as shown in the following screenshot:

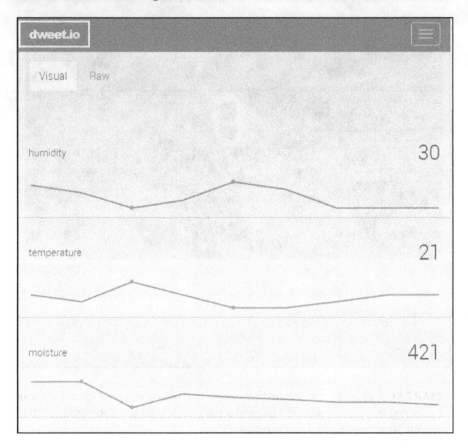

The disadvantage of using the `dweet.io` graphical representation of posted data is that you do not have any other options for how to present your data other than the line graph. For a more comprehensive dashboard, try `freeboard.io`:

1. When using `freeboard.io`, the first thing to do is to create a new dashboard by entering the name of the new dashboard and clicking on the **Create New** button on the `freeboard.io` home page. We will call our dashboard `garden-monitor`:

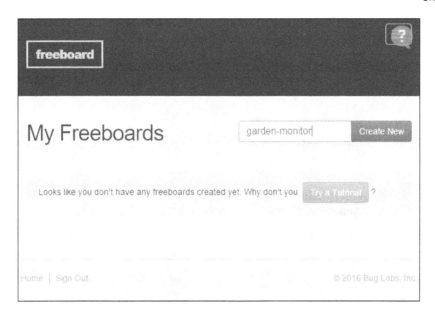

2. A new empty dashboard will be created and will look like the following screenshot:

3. Proceed to add a datasource by clicking on the **ADD** link under the **DATASOURCES** section. This should open up a window that looks like this:

4. Use this window to select a **Type** of data source. Set the name of the datasource and the thing name, and enter the special key if the thing is locked. This is how our datasource setup looks:

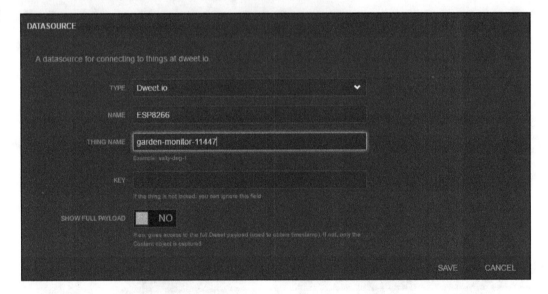

5. Once you are done with the setup, click on **SAVE**. The new datasource will be listed in the **DATASOURCE** section.

6. The next thing to do is to create a widget that we will use to display our data. To do that, click on the **ADD PANE** link to create a new pane. The pane will be added to the dashboard, as shown in the following screenshot:

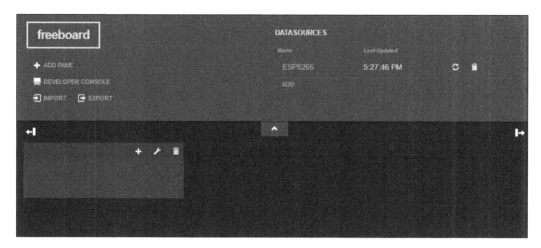

7. To create a widget, click on the **+** icon in the newly created pane. This will open up the widget setup window. The first thing you choose is the type of widget you are creating; since we are setting up a widget for temperature, we'll go with a GAUGE widget. The **WIDGET** window with the gauge setup is shown in the following screenshot:

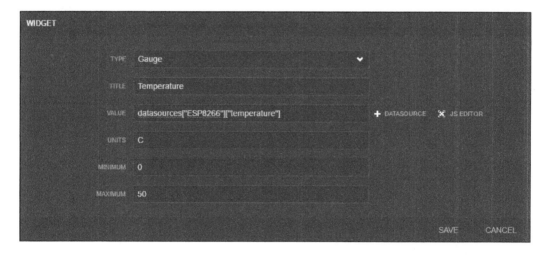

8. To select the datasource in the **WIDGET** window, click the **DATASOURCE** button next to the **VALUE** text box. It lists the available data sources and parameters, so you can select the ones that you want to use on the widget.

9. When you are satisfied with your setup, click on **SAVE** and wait for a few seconds. If there is temperature data already posted on `dweet.io`, a **GAUGE** will appear and visualize the last updated data, as shown in the screenshot:

10. You can now repeat the process to create widgets to visualize the `humidity` and `soil moisture` data. Our final dashboard looks like this:

Try out several other widgets and explore the different things that you can do with them. Once done, you will be able to create a comprehensive dashboard that you can use to monitor all your sensor data.

See also

When using `dweet.io`, you can generate automated alerts to let you know when some sensor value has exceeded or gone beyond a certain threshold. Proceed to the next recipe to learn how to do it.

Creating automated alerts based on the measured data

In this recipe, we will be looking at how to create alerts based on the data you logged online. Alerts are available for locked dweets. They send notifications to you when the posted data exceeds a certain limit. This is an important feature for real-time monitoring. To demonstrate this, we will create an alert to inform us when the temperature exceeds 25 degrees Celsius.

Getting ready

You will need an ESP8266 board, a USB cable, and several other hardware components:

- DHT11 (`https://www.sparkfun.com/products/10167`)
- Soil moisture sensor (`https://www.sparkfun.com/products/13322`)
- 220 Ω resistor
- 100 Ω resistor
- 10 kΩ resistor
- Breadboard
- Jumper wires

Set up the hardware as we did in the, *Connecting sensors to your ESP8266 board recipe*:

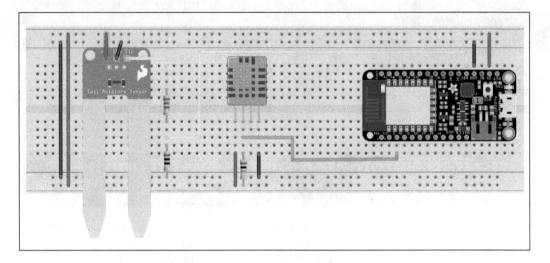

We will still use same thing name, `garden-monitor-11447`, when posting data online. However, you may need to use another thing name just in case the one suggested earlier is locked. If that is the case, remember to use your new thing name in all the URLs.

Since alerts work with locked things, it is advisable to get yourself a lock and key before proceeding to the next section. Visit `https://dweet.io/locks` to do that.

How to do it...

1. To create an alert, make an `http` request using a URL with this format: `https://dweet.io/alert/{recipients}/when/{thing}/{condition}?key={key}`

   In the preceding URL:

   - `{recipients}`: This refers to e-mail addresses you want `dweet.io` to send notifications to. If there are several e-mail addresses, separate them using commas.
   - `{thing}`: This is the valid name for your thing.
   - `{condition}`: This is a JavaScript expression that you use to evaluate data stored in `dweet.io`. For example: `if(dweet.temp <= 32) return "frozen"; else if(dweet.temp >= 212) return "boiling";`

   The expression is limited to 2,000 characters and should not contain JavaScript reserved words, loops, or other complex things:

   - `{key}`: This is the valid key for your locked dweet

2. To create an `alert` when the temperature exceeds 25 degrees Celsius, we will use this URL: `https://dweet.io/alert/my-email@domain.com/when/garden-monitor-11447/if(dweet.temperature > 25) return "Too hot";?key={key}`.

   ❑ You can remove alerts using this URL format:

   `https://dweet.io/remove/alert/for/{thing}?key={key}`

   ❑ Instead of using your ESP8266 board to lock your thing and create an `alert`, you can use a web browser. Using the web browser is the fastest way to do it, since all you need to do is call the URL for locking dweets and the URL for setting alerts on your Internet browser. Moreover, you will only need to lock your thing and create an alert once, so there is no need for your ESP8266 board to repeat the process every time it turns on.

   ❑ Therefore, use your web browser to call this URL to lock your thing: `https://dweet.io/lock/garden-monitor-11447?lock={your_lock}&key={your_key}`.

3. Replace `{your_lock}` with the lock ID you were sent via e-mail and replace `{your_key}` with the master key that was provided in the e-mail.

   ❑ Once the thing has been locked successfully, create an alert to send you notifications when the temperature exceeds 25 degrees Celsius. Do that by calling this URL in your web browser: `https://dweet.io/alert/my-email@domain.com/when/garden-monitor-11447/if(dweet.temperature > 25) return "Too hot";?key={key}`.

4. Remember to replace `{key}` with your master key.

5. After using your web browser to lock your thing and create an alert, upload the data logging sketch to your ESP8266. This will log sensor data on `dweet.io` and inform you when the temperature goes above 25 degrees Celsius:

```
// Libraries
#include <ESP8266WiFi.h>
#include "DHT.h"

// create wifi client object


// Wi-Fi network SSID and password
const char* ssid     = "your-ssid";
const char* password = "your-password";

// Host
const char* host = "dweet.io";

// dweet.io lock and key
```

```
String key = "your-key";

#define DHTPIN 2  // what digital pin DHT11 is connected to
#define DHTTYPE DHT11 // DHT 11 sensor

DHT dht(DHTPIN, DHTTYPE);

int moistureReading = 0; // holds value soil moisture sensor
reading

void setup() {
  Serial.begin(115200); // initialize serial interface
  dht.begin(); // initialize DHT11 sensor
  delay(10);

  // We start by connecting to a WiFi network
  Serial.println();
  Serial.println();
  Serial.print("Connecting to ");
  Serial.println(ssid);

  WiFi.begin(ssid, password);

  while (WiFi.status() != WL_CONNECTED) {
    delay(500);
    Serial.print(".");
  }

  Serial.println("");
  Serial.println("WiFi connected");
  Serial.println("IP address: ");
  Serial.println(WiFi.localIP());

}

void loop() {
  delay(5000);

  Serial.print("connecting to ");
  Serial.println(host);

  WiFiClient client;
  const int httpPort = 80;
  // Use WiFiClient class to create TCP connections
```

```
if (!client.connect(host, httpPort)) {
  Serial.println("connection failed");
  return;
}

// Read sensor inputs
// get humidity reading
float h = dht.readHumidity();
// get temperature reading in Celsius
float t = dht.readTemperature();
// Check if any reads failed and exit early (to try again).
while (isnan(h) || isnan(t)) {
  Serial.println("Failed to read from DHT sensor!");
  delay(2000); // delay before next measurements
  //get the measurements once more
  h = dht.readHumidity();
  t = dht.readTemperature();
}
//get soil moisture reading
moistureReading = analogRead(A0);


// We now create a URI for the request
String url = "/dweet/for/garden-monitor-11447?key=";
url += key;
url += "&humidity=";
url += String(h);
url += "&temperature=";
url += String(t);
url += "&moisture=";
url += String(moistureReading);

// Send request
Serial.print("Requesting URL: ");
Serial.println(url);

client.print(String("GET ") + url + " HTTP/1.1\r\n" +
             "Host: " + host + "\r\n" +
             "Connection: close\r\n\r\n");
unsigned long timeout = millis();
while (client.available() == 0) {
  if (millis() - timeout > 5000) {
    Serial.println(">>> Client Timeout !");
    client.stop();
```

```
            return;
        }
    }
    // Read all the lines from the answer
    while(client.available()){
        String line = client.readStringUntil('\r');
        Serial.print(line);
    }

    // Close connecting
    Serial.println();
    Serial.println("closing connection");
}
```

- ❑ This sketch is the same as the one for logging data to an unlocked thing on dweet.io. The only difference is that this time the key is included in the URL, so you will have to provide the master key in this section of the code:

```
// dweet.io lock and key
String key = "your-key";
```

- ❑ The key is then appended to the URL in this section of the code:

```
// We now create a URI for the request
    String url = "/dweet/for/garden-monitor-11447?key=";
    url += key;
```

6. Copy the sketch to your Arduino IDE and change the `ssid` in the code from your-ssid to the name of your Wi-Fi network, and change the `password` from your-password to the password of your Wi-Fi network. Also, change the `key` from your-key to the key that was provided to you via e-mail. Use the master key, not the read-only key.

7. Upload the sketch to your ESP8266 board and open the serial monitor so that you can view the incoming data.

How it works...

The program connects to the Wi-Fi network using the provided `password` and `ssid`. It then proceeds to connect to the provided cloud/host server using the `client.connect()` function. Once the ESP8266 connects successfully, data from the sensors is then read and a URL is generated that includes the updated sensor data and the master key for the lock. The URL is then sent to the `host` server using the `client.print()` function.

Once the data has been successfully sent, the sketch waits for a reply from the server. It does this with the `client.available()` function, which checks whether there is incoming data from the server. If there is data available, the sketch reads it and displays it on the serial monitor. The ESP8266 posts sensor data to `dweet.io` every five seconds.

`dweet.io` monitors the temperature value and sends alerts to your e-mail address, when the temperature exceeds 25 degrees Celsius. The e-mail will look like this:

An alert has been OPENED for the thing 'garden-monitor-11447'.

It said: Too hot

There's more...

Try and change the current alert to be triggered when the temperature exceeds another value. You can even add alerts for the other parameters, `humidity` and `soil moisture`.

See also

Now that you can successfully monitor data from one ESP8266 board, you can take it a notch higher and start monitoring more than one ESP8266 module at the same time. The next recipe will show you how to do that.

Monitoring several ESP8266 modules at once

In advanced and more complex monitoring solutions, you will have to use several ESP8266 modules. This may be due to the distance between the areas where parameters are being monitored, or when the sensors being read are more than the number of GPIO pins available on an ESP8266 module. In such cases, you will have to monitor data from more than one ESP8266 module on the same dashboard. To demonstrate how to do that, we will use one ESP8266 board to monitor temperature and humidity with the DHT11 sensor, and another ESP8266 board to monitor soil moisture. Data from both ESP8266 boards will be monitored using one dashboard.

Getting ready

You will need these hardware components:

- ▶ Two ESP8266 boards
- ▶ Two USB cables
- ▶ DHT11 (https://www.sparkfun.com/products/10167)
- ▶ Soil moisture sensor (https://www.sparkfun.com/products/13322)
- ▶ 220 Ω resistor
- ▶ 100 Ω resistor
- ▶ 10 kΩ resistor
- ▶ Breadboard
- ▶ Jumper wires

Start by mounting the ESP8266 board and the DHT11 sensor onto the breadboard. Then connect a **10** kΩ pull up resistor to the DHT11 data pin and connect the VCC pin and **GND** pin of the sensor to the 3V pin and **GND** pin of the ESP8266 board, respectively. Finally, connect the data pin of the DHT11 to GPIO 2 of the ESP8266 board.

The first setup will look like this:

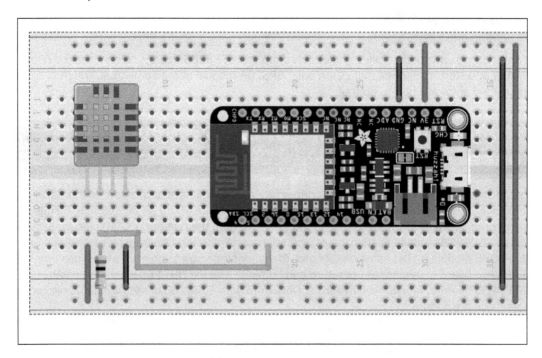

For the second setup, begin by connecting the soil moisture sensor **VCC** and **GND** pins to the ESP8266 board 3V and **GND** pins respectively. Then connect the **SIG** pin to the voltage divider. The voltage divider will be constructed using the 220 Ω and 100 Ω resistors. Connect the output of the voltage divider to the analog pin.

The setup will look like this:

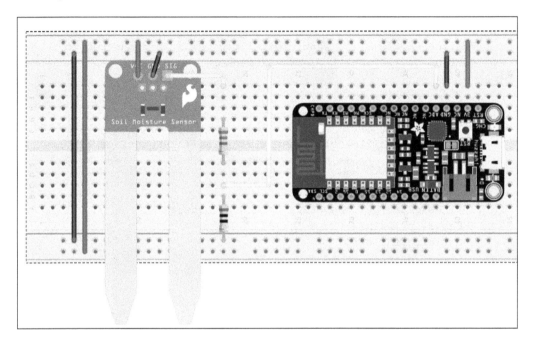

How to do it...

1. Create two dweets using different thing names. The names we chose are `garden-monitor-11447` and `garden-monitor-11448`.

2. The first thing name is for the ESP8266 board that has the `DHT11` sensor attached to it, while the second one is for the ESP8266 board that monitors soil moisture.

3. The URLs you will use to send the dweets will look like this: `https://dweet.io/dweet/for/garden-monitor-11447?humidity={humidity value}&temperature={temperature value}` and `https://dweet.io/dweet/for/garden-monitor-11448?moisture={moisture value}`

4. The sketch for the ESP8266 board that monitors temperature and humidity is as follows:

```
// Libraries
#include <ESP8266WiFi.h>
#include "DHT.h"
```

```
// Wi-Fi network SSID and password
const char* ssid     = "your-ssid";
const char* password = "your-password";


// Host
const char* host = "dweet.io";


#define DHTPIN 2  // what digital pin DHT11 is connected to
#define DHTTYPE DHT11 // DHT 11 sensor

DHT dht(DHTPIN, DHTTYPE);

void setup() {
  Serial.begin(115200); // initialize serial interface
  dht.begin(); // initialize DHT11 sensor
  delay(10);

  // We start by connecting to a WiFi network
  Serial.println();
  Serial.println();
  Serial.print("Connecting to ");
  Serial.println(ssid);

  WiFi.begin(ssid, password);

  while (WiFi.status() != WL_CONNECTED) {
    delay(500);
    Serial.print(".");
  }

  Serial.println("");
  Serial.println("WiFi connected");
  Serial.println("IP address: ");
  Serial.println(WiFi.localIP());
}

void loop() {
  delay(5000);

  Serial.print("connecting to ");
  Serial.println(host);
  WiFiClient client;
  const int httpPort = 80;
```

```
// Use WiFiClient class to create TCP connections
if (!client.connect(host, httpPort)) {
  Serial.println("connection failed");
  return;
}


// Read sensor inputs
// get humidity reading
float h = dht.readHumidity();
// get temperature reading in Celsius
float t = dht.readTemperature();
// Check if any reads failed and exit early (to try again).
while (isnan(h) || isnan(t)) {
  Serial.println("Failed to read from DHT sensor!");
  delay(2000); // delay before next measurements
  //get the measurements once more
  h = dht.readHumidity();
  t = dht.readTemperature();
}



// We now create a URI for the request
String url = "/dweet/for/garden-monitor-11447?humidity=";
url += String(h);
url += "&temperature=";
url += String(t);

// Send request
Serial.print("Requesting URL: ");
Serial.println(url);

client.print(String("GET ") + url + " HTTP/1.1\r\n" +
             "Host: " + host + "\r\n" +
             "Connection: close\r\n\r\n");
unsigned long timeout = millis();
while (client.available() == 0) {
  if (millis() - timeout > 5000) {
    Serial.println(">>> Client Timeout !");
    client.stop();
    return;
  }
}
// Read all the lines from the answer
while(client.available()){
```

```
      String line = client.readStringUntil('\r');
      Serial.print(line);
    }

    // Close connecting
    Serial.println();
    Serial.println("closing connection");
  }
```

5. The sketch for the ESP8266 board that monitors the soil moisture is as follows:

```
// Libraries
#include <ESP8266WiFi.h>

// Wi-Fi network SSID and password
const char* ssid     = "your-ssid";
const char* password = "your-password";

// Host
const char* host = "dweet.io";

int moistureReading = 0; // holds value soil moisture sensor
reading

void setup() {
  Serial.begin(115200); // initialize serial interface
  delay(10);

  // We start by connecting to a WiFi network
  Serial.println();
  Serial.println();
  Serial.print("Connecting to ");
  Serial.println(ssid);

  WiFi.begin(ssid, password);

  while (WiFi.status() != WL_CONNECTED) {
    delay(500);
    Serial.print(".");
  }

  Serial.println("");
  Serial.println("WiFi connected");
  Serial.println("IP address: ");
  Serial.println(WiFi.localIP());
}
```

```
void loop() {
  delay(5000);

  Serial.print("connecting to ");
  Serial.println(host);

  WiFiClient client;
  const int httpPort = 80;
  // Use WiFiClient class to create TCP connections
  if (!client.connect(host, httpPort)) {
    Serial.println("connection failed");
    return;
  }

  //get soil moisture reading
  moistureReading = analogRead(A0);


  // We now create a URI for the request
  String url = "/dweet/for/garden-monitor-11448?moisture=";
  url += String(moistureReading);

  // Send request
  Serial.print("Requesting URL: ");
  Serial.println(url);

  client.print(String("GET ") + url + " HTTP/1.1\r\n" +
               "Host: " + host + "\r\n" +
               "Connection: close\r\n\r\n");
  unsigned long timeout = millis();
  while (client.available() == 0) {
    if (millis() - timeout > 5000) {
      Serial.println(">>> Client Timeout !");
      client.stop();
      return;
    }
  }
  // Read all the lines from the answer
  while(client.available()){
    String line = client.readStringUntil('\r');
    Serial.print(line);
  }

  // Close connecting
  Serial.println();
  Serial.println("closing connection");
}
```

Remember to change the `password` and `ssid` in your code so that they match those of your Wi-Fi network. Then upload the codes to their respective ESP8266 boards. Once that is done, the sensor data will be posted to `dweet.io` and now you can proceed to monitor the data on `freeboard.io`.

Since you have already created an account and a dashboard on `freeboard.io` in the, *Monitoring sensor data from a cloud dashboard* recipe, all you are required to do now is to add another datasource to your dashboard for the `garden-monitor-11448` thing. This will enable us to view soil moisture data from the second ESP8266 board on our dashboard.

The first thing you should do is delete the soil moisture widget and pane that you created earlier.

Then you can add a **Sparkline** widget to the **Humidity** pane so that you can also monitor the humidity trends. To do this, follow these steps:

1. Click on the **+** icon on the **Humidity** pane, which is highlighted in the following screenshot:

2. It will bring up the **WIDGET** creation window. Select a **Sparkline**, as shown in the following screenshot:

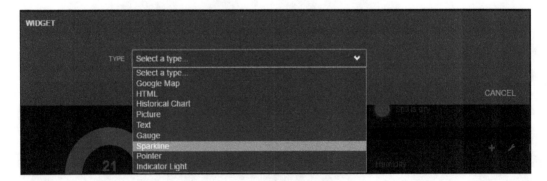

3. Enter the **Sparkline** settings as shown:

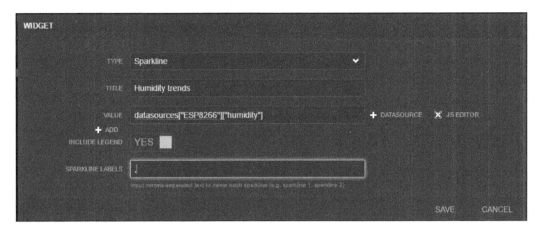

4. When you click on **SAVE**, the **Humidity** widget will appear as shown in the following screenshot. You can also do the same to the **Temperature** widget as was done in our case.

5. Now you can monitor the temperature and humidity from the first ESP8266, using the datasource we created in an earlier tutorial:

6. To monitor the soil moisture levels on this dashboard, you will have to add another datasource that is linked to the `garden-monitor-11448` thing on `dweet.io`. To do that, click on the **ADD** link in the **DATASOURCE** area of your dashboard:

7. You then set up a new datasource. We called ours `ESP8266-2` and linked it to the `garden-monitor-11448` thing on `Dweet.io`. The settings look like this:

8. Clicking **SAVE** will create the new datasource.

9. You can then create a new pane by clicking on the **Add Pane link** then your dashboard, and then clicking on the **+** on the new pane so that you can create a new widget for soil moisture monitoring.

10. First, create a **Gauge** widget for soil moisture. The settings will look like this:

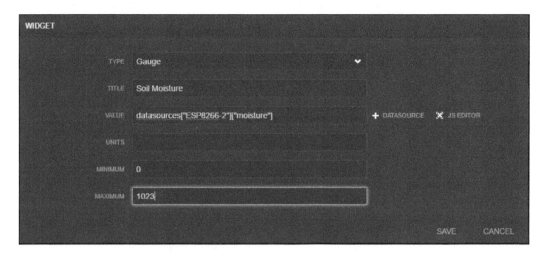

11. Then, create a **Sparkline** widget in the same pane with the following settings:

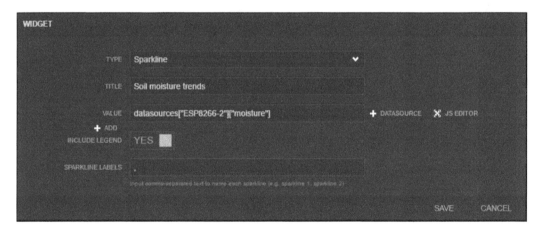

Notice that we use the `ESP8266-2` datasource in both widgets.

Once that is done, your dashboard will be complete. You will be able to monitor all the parameters, **Temperature**, **Humidity**, and **Soil Moisture**, from two ESP8266 modules using one dashboard:

How it works...

The ESP8266 board with a DHT11 connected to it measures the temperature and humidity, and posts it to dweet.io using the thing name garden-monitor-11447. The other ESP8266 board posts analog readings from the soil moisture sensor to dweet.io using the thing name garden-monitor-11448.

freeboard.io uses two data sources: the **ESP8266** datasource, which is connected to the garden-monitor-11447 thing, and ESP8266-2 datasource, which is connected to the garden-monitor-11448 thing. Using these two datasource, the widgets can acquire temperature, humidity, and soil moisture readings from dweet.io and display them on the dashboard.

There's more...

Try and use other widgets to display data on the dashboard.

See also

There are some common problems that you may face when using web services. The next recipe will describe them and show you how to solve them.

Troubleshooting common issues with web services

In this recipe, we will discuss some of the problems you may run into and how to troubleshoot and solve them.

The board is not connecting to the Wi-Fi network

This usually happens if the Wi-Fi `ssid` and `password` provided in the code do not match those of the Wi-Fi network that the ESP8266 is supposed to connect to. You can solve this by providing the correct credentials in your code.

The lock feature on dweet is not working

The lock feature fails to work if the lock is already being used on another thing. On the other hand, it can fail if the lock ID and key are not entered correctly. Therefore, first check you have entered the correct lock ID and the correct key. If they are already correct, check whether the lock is being used on another thing. If another thing is using the lock, just unlock it using this URL: `https://dweet.io/unlock/{thing_name}?key={your_key}`

If you are not sure of the name of the other locked thing, call this URL so that you unlock anything that is locked by your lock:

`https://dweet.io/remove/lock/{your_lock}?key={your_key}`

Then you can lock the new thing.

The alert feature on dweet is not working

There are two possible reasons for this: the first one is that you are trying to add an alert to a thing that is not locked; the second reason is that the condition in your `http` request is not well written. So, make sure that the thing is locked and that the condition, in your URL, is a simple JavaScript expression.

The widgets on my dashboard do not display readings

This can happen if the data your thing is posting to `dweet.io` is not numerical. If the parameter value is not in numerical format, `freeboard.io` widgets cannot graphically present the data saved on `dweet.io`. Therefore, ensure that your thing is posting numerical parameter values.

If your thing is already posting numerical data, but your widgets are not showing anything, check the widget settings to confirm that they are set up in the right way.

6
Interacting with Web Services

In this chapter, we will cover:

- ► Discovering the Temboo platform
- ► Tweeting data from the Arduino board
- ► Posting updates on Facebook
- ► Storing data on Google Drive
- ► Automation with IFTTT
- ► Sending push notifications
- ► Sending e-mail notifications
- ► Sending text message notifications
- ► Troubleshooting common issues with web services

Introduction

One of the main functions of the IoT is to provide instant notifications from remotely located devices. Online dashboards are able to facilitate instant data monitoring, but are not very helpful in real-time monitoring on the go. This is because some dashboards are not compatible with phones and require an Internet connection.

It is for this reason that IoT platforms are embracing social media platforms, e-mail, and text messages, to provide instant notifications and updates from IoT devices. Most phones are configured to receive notifications when there are updates on social media platforms or when new e-mails are received. This way, people can receive instant alerts on their Internet-enabled phones. Text messages, on the other hand, are considered to be the best way to send alerts from IoT devices. This is because the alerts can be received by all types of phones without requiring an active Internet connection.

In this chapter, we will look at how to send tweets, Facebook updates, push notifications, text messages, and e-mails from the ESP8266. This will provide you with enough knowledge on how to get notifications from your ESP8266-based IoT devices.

Discovering the Temboo platform

Temboo is a fully fledged online IoT platform that links your IoT projects to numerous other online platforms for instant notifications, data logging, and control purposes. You can use Temboo with many open-source hardware IoT platforms. In this chapter, we will be looking at using Temboo with the Arduino platform, and the ESP8266. Note that Temboo is not free, but you can make several requests to the service with their free plan.

There are many functions of the Temboo platform that come in handy in IoT projects. Some of them include the following:

- **Code generation**: The platform allows you to select the device and sensors you want to use and automatically generates code that you can use to achieve your desired task

- **Sensor data monitoring**: You are able to monitor your sensors using graphical means and keep track of your IoT systems

- **Cloud controls**: The platform provides an interface that you can use to control your actuators with just one click

Temboo has a comprehensive library that simplifies IoT programs. Using the library, you can run any process with just a few lines of code. In addition to providing an Arduino-compatible library, Temboo generates code on the website, so you only have to copy and paste it on the IDE you are using. This makes it easy for you to use Choreos.

Choreos are the smart processes provided by Temboo. When you run a sketch that has Choreos, it calls the Temboo platform, where the more complicated code that makes up the Choreo is executed. This is called code virtualization. It takes the processing load off the microcontroller, which has less processing power, and offloads it to powerful Temboo servers. There are thousands of Choreos for numerous functions, such as working databases, performing code utility functions, and for API interactions, among many other things. Some of the online platforms whose Choreos are available include the following:

- Amazon
- Facebook

- ▶ GitHub
- ▶ Dropbox
- ▶ Google
- ▶ Paypal
- ▶ Twitter
- ▶ Ebay

There are several advantages of using Temboo in your IoT projects. First of all, it relieves the microcontroller of the major processing activities. Therefore, you can increase the functionality of your IoT project, since there are more resources to work with. Temboo also reduces the amount of microcontroller code you have to write to interact with an online resource. You can now accomplish so much with just a few lines of code.

Since Temboo works with different IoT hardware platforms, it is easy for you to migrate your code to other boards if need be. This way you are able to create versatile IoT projects that you can improve on over time, using better and more powerful hardware.

To use Temboo, you need to create a Temboo account. You are required to provide your credentials, as shown in the following form:

Once you have an account, you can start using Temboo in your IoT projects and discover the endless possibilities it offers.

See also

Now that you have a rough idea of how Temboo works, you can go to the next recipe and learn how to send tweets from your **ESP8266** board via Temboo.

Tweeting data from the ESP8266 board

This recipe will look at how to tweet sensor data directly from the ESP8266 board. To show you how to do it, we are going to use an ESP8266 board to tweet temperature readings from the DHT11 sensor, through the Temboo platform.

Getting ready

The hardware setup will require the following components:

- ESP8266 board
- USB cable
- 1 kΩ - 10 kΩ photocell (https://www.sparkfun.com/products/9088)
- 1 kΩ resistor
- Breadboard
- Jumper wires

Mount the ESP8266 board onto the breadboard. Then connect one leg of the photocell to the 3V pin of the ESP8266 board and the other leg to the 1 kΩ resistor. Connect the 1 kΩ resistor's other leg to the **GND** pin of the ESP8266 board, so that the photocell and resistor make a voltage divider circuit. Then connect the junction of the photocell and resistor to the ADC pin of the ESP8266 board.

The setup will look like this:

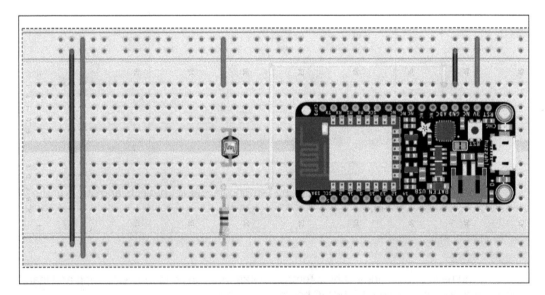

How to do it...

Refer to the following steps:

1. Log in to your Temboo account. The first page that you will land on will be the libraries page. It shows if you have any Choreos running and provides a graph that displays how many runs have been made in the current month, as well as their success rate.

2. Since you want to send tweets from your ESP8266 board, select **Twitter | Tweets** from the libraries menu on the left side of the screen. This directs you to the Tweets Choreos page, where you will be able to register your application with Twitter and get the API keys that you need when setting up your Choreos.

3. Click on the **register** link to open the Twitter app registration form. You will be required to provide the following things:

   ❏ Name of your application

   ❏ Description of your application

   ❏ Website URL, if you have one

   ❏ Callback URL, if need be

4. Once the app is successfully created, you will get your API key and API secret from the **Keys and Access Tokens** tab. You will use them as the `ConsumerKey` and `ConsumerSecret` respectively, when setting up your Choreo.

5. To create the `AccessToken` and the access token secret, click on the **Create my access token** button in the **API Key** tab shown in the following screenshot:

Normally, Twitter apps have read only permission. You have to change this if you want your application to successfully post data on Twitter:

1. To do that, go to the **Permissions** tab on the application page and select **Read and Write**.

2. Click on **Update Settings** button to apply the changes to your application. Once that is done, regenerate your `AccessTokens` on the **Keys and Access Tokens** tab.

3. Now that you have all the authentication keys you need, click on the **StatusUpdate Choreo** on the Tweets page. It will open the status update page, which provides you with everything you need to design your program. Start by selecting your device and how it is connected:

We chose the Arduino 101 as our device, since Temboo does not support ESP8266 directly. The connection was set as Arduino Wi-Fi 101 shield, which we named ESP8266:

1. Once you have set your device and connection, proceed to input `AccessToken`, `AccessTokenSecret`, `ConsumerKey`, `ConsumerSecret`, and `StatusUpdate`, in the input section of the page. You will get these authentication keys from the Twitter application you created. `StatusUpdate` is the tweet that is going to be sent.

2. The next thing to do is to select the pin that the LDR sensor output will be connected to. In our case, it is **A0**:

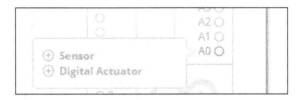

3. Click on **Sensor** and specify that you are sensing light. Select the type of light sensor as **Other** and provide its name. We called ours **LDR**. You can then proceed to set the condition that will trigger an alert on Twitter and the time interval that the condition should be checked:

4. Now you can click on **Generate Code** at the end of the input section to test the Choreo with the current settings. You can access the automatically generated code in the code section, while the `header` file section has all the Temboo API keys and credentials, as well as Wi-Fi network credentials.

5. To download the code and `header` file, click on the **Download source code** link. A `zip` file that contains the Arduino sketch and `header` file will be downloaded. You have to edit the code a little before uploading it onto your ESP8266 board. This is because the code that was generated was meant for the Arduino 101, with an Arduino Wi-Fi 101 shield.

6. The changes you need to make to the code are as follows:

   ❑ Replace the included libraries in the Arduino sketch with the `ESP8266` libraries

   ❑ Provide your Wi-Fi network credentials in the `header` file

   ❑ Replace the code in the setup section with the one you will use for connecting the ESP8266 board to a Wi-Fi hotspot

   ❑ The code will look like this after editing. Include the `Temboo` and `ESP8266` libraries:

   ```
   #include <SPI.h>
   #include <ESP8266WiFi.h>
   #include <WiFiClient.h>
   #include <Temboo.h>
   #include "TembooAccount.h" // Contains Temboo account
   information
   ```

   ❑ Create the `WiFiClient` object and initialize the variables that are going to be used in the sketch:

   ```
   WiFiClient client;

   // The number of times to trigger the action if the condition is
   met
   // We limit this so you won't use all of your Temboo calls while
   testing
   int maxCalls = 10;

   // The number of times this Choreo has been run so far in this
   sketch
   int calls = 0;

   // Choreo execution interval in milliseconds
   unsigned long choreoInterval = 300000;
   // Store the time of the last Choreo execution
   unsigned long lastChoreoRunTime = millis() - choreoInterval;
   ```

   ❑ Define the pin that the `sensor` will be connected to:

   ```
   // Declaring sensor configs
   TembooGPIOConfig sensorA0Config;
   // Declaring TembooSensors
   TembooSensor sensorA0;
   ```

❏ Connect the ESP8266 module to the Wi-Fi network:

```
void setup() {
  Serial.begin(115200);
  Serial.print("Connecting to Wi-Fi hotspot");

  WiFi.begin(WIFI_SSID, WPA_PASSWORD);
  while (WiFi.status() != WL_CONNECTED) {
    delay(500);
    Serial.print(".");
  }
  Serial.println("");
  Serial.println("WiFi connected");
  Serial.println("IP address: ");
  Serial.println(WiFi.localIP());
  delay(1000);
}
```

❏ Read sensor data and check if the condition has been met, then execute the Choreo. This is done every 5 minutes:

```
void loop() {
  if(millis() - lastChoreoRunTime >= choreoInterval) {
    //read sensor value
                  int sensorValue = sensorA0.read(sensorA0.
sensorConfig);
    Serial.println("Sensor: " + String(sensorValue));
                  //check if sensor value is less than 614
    if (sensorValue < 614) {
      if (calls < maxCalls) {
        Serial.println("\nTriggered! Calling StatusesUpdate
Choreo...");
        runStatusesUpdate(sensorValue); // run Choreo
        lastChoreoRunTime = millis(); // record last time choreo
was run
        calls++;
      } else {
        Serial.println("\nTriggered! Skipping to save Temboo
calls. Adjust maxCalls as required.");
      }
    }
  }

  if (millis() - lastChoreoRunTime >= ULONG_MAX - 10000) {
    lastChoreoRunTime = millis() - choreoInterval;
  }
  delay(250);
}
```

Configures StatusUpdateChoreo and runs it:

```
void runStatusesUpdate(int sensorValue) {
  TembooChoreo StatusesUpdateChoreo(client);

  // Set Temboo account credentials
  StatusesUpdateChoreo.setAccountName(TEMBOO_ACCOUNT);
  StatusesUpdateChoreo.setAppKeyName(TEMBOO_APP_KEY_NAME);
  StatusesUpdateChoreo.setAppKey(TEMBOO_APP_KEY);

  // Set profile to use for execution
  StatusesUpdateChoreo.setProfile("esplog");
  // Set Choreo inputs
  String StatusUpdateValue = "Its Dark";
  StatusesUpdateChoreo.addInput("StatusUpdate",
StatusUpdateValue);
  // Identify the Choreo to run
  StatusesUpdateChoreo.setChoreo("/Library/Twitter/Tweets/
StatusesUpdate");

  // Run the Choreo
  unsigned int returnCode = StatusesUpdateChoreo.run();

  // Read and print the error message
  while (StatusesUpdateChoreo.available()) {
    char c = StatusesUpdateChoreo.read();
    Serial.print(c);
  }
  Serial.println();
  StatusesUpdateChoreo.close();
}
```

❑ The header file is as follows:

```
#define TEMBOO_ACCO-UNT "yhtomit"  // Your Temboo account name
#define TEMBOO_APP_KEY_NAME "myFirstApp"  // Your Temboo app key
name
#define TEMBOO_APP_KEY "d7EWkxxxxxxxxxxxxxxxxxxxxxxxx"  // Your
Temboo app key
#define TEMBOO_DEVICE_TYPE "a101+w101"

#define WIFI_SSID "ssid"
#define WPA_PASSWORD "pass"

#if TEMBOO_LIBRARY_VERSION < 2
#error "Your Temboo library is not up to date. You can update it
using the Arduino library manager under Sketch > Include Library >
Manage Libraries..."
#endif
```

The `Temboo` library that comes with the Arduino IDE does not support ESP8266 boards, and will probably bring up compiling errors when you try to upload it. To solve this, download an updated version of the `Temboo` library that works with ESP8266 from this link `https://github.com/marcoschwartz/esp8266-IoT-cookbook`.

To install the library on your computer, you can check this guide:

`https://www.arduino.cc/en/guide/libraries`.

1. Change the SSID in the `header` file from `ssid` to the name of your Wi-Fi network and the password from `password` to the password of your Wi-Fi network.
2. Copy the sketch and paste it in your Arduino IDE. Check to ensure that the ESP8266 board is connected.
3. Select the board that you are using in **Tools | Board menu** (in this case, it is the **Adafruit HUZZAH ESP8266**).
4. Select the serial port your board is connected to from the **Tools | Port menu** and then upload the code.

You can open the serial monitor to view status messages and replies from the Twitter servers.

How it works...

The program starts by including the `ESP8266` libraries and the `Temboo` library, and creates a `WiFiClient` object. The sketch then connects to the Wi-Fi hotspot whose credentials you have provided. Once connected, the sketch reads the LDR sensor output every 5 minutes. If the analog reading goes down by 614, the `StatusUpdate` Choreo is called and a tweet is posted on your Twitter account. The tweet will read `its dark`.

The number of times the StatusUpdate Choreo is called is tracked, and when 10 tweets have been sent, the sketch no longer sends tweets when the sensor analog output goes down by 614.

There's more...

Increase or reduce the time interval it takes for the sensor value to be read. It is currently set at 5 minutes. You can also change the condition so that the StatusUpdate Choreo gets called when the sensor analog output goes above 0.5V (approximately 512).

See also

Now that you have managed to successfully tweet data from your ESP8266 board, you should be able to do the same with Facebook. The next recipe will show you how to do it.

Posting updates on Facebook

The ESP8266 board can post Facebook updates. With this capability, you can receive real-time data from your IoT projects wherever you are. While this sounds like a difficult task, it is actually easy when using Temboo, thanks to its Facebook Choreo. This recipe will show you how to configure the Choreo to successfully post updates on Facebook.

Getting ready

For this recipe, you will need an ESP8266 board, a USB cable, and a Temboo account. If you do not have a Temboo account, check the first recipe in this chapter to see how you can get one.

How to do it...

Refer to the following steps:

1. Navigate to **Facebook** | **Publishing** on the libraries menu on Temboo. This will open the Facebook publishing page.

2. Then sign up for a Facebook developer account if you do not have one already. The link is provided in the setup instructions on the publishing page.

3. Once you have a Facebook developer account, proceed to create an app in the developer console. Use the **My Apps** menu at the top of the page to do that. In the platform option, select **Website** and enter the site URL as http://temboo.com.

4. Enter: https://{ACCOUNT_NAME}.tembovlive.com/callback/ in the Valid Auth redirect URIs field under the **Facebook Login** | **Settings** in the **Client OAuth Settings** section. Replace {ACCOUNT_NAME} with your Temboo account name:

5. Now that you have created the app successfully, select the **Facebook** | **OAuth** item in the Temboo libraries menu. Choose the **IntializeOAuth Choreo** from the OAuth page. This Choreo will enable you to acquire the AuthorizationURL and **CallbackID** that you will use to give your app permission to publish items on Facebook and to get the access token, respectively.

6. In the `InitializeOAuth` Choreo, provide the AppID from your Facebook app dashboard and specify the permissions to request access for in the `Scope` field. In this recipe, we used `publish_actions` as the scope.

7. Once that is done, click on **Generate Code**. Copy the AuthorizationURL and **CallbackID**, or just leave the webpage open, since you will need them in the next step.

8. Open the AuthorizationURL on a separate tab on your web browser. The webpage will enable you to grant your app access to Facebook.

9. After that, open the `FinalizeOAuth` Choreo from the OAuth page. Provide the AppID and AppSecret of your Facebook app in the Choreo. You can find them in the **Facebook** app dashboard. Also provide the **CallbackID** generated by the InitializeOAuth Choreo.

10. Click on **Generate Code** so that the FinalizeOAuth Choreo can generate an access token.

11. Now that you have the access token, proceed to the `SetStatusChoreo` and generate a sketch that you will use to post status updates from your ESP8266 board.

12. Open the `SetStatus Choreo` from the Facebook publishing page.

13. Then select the Arduino board you are using, which in our case is the **Arduino101**, and the Arduino Wi-Fi 101 shield, which we named `ESP8266` in the previous recipe:

14. Copy and paste the access token to the input section and write the status update you want your ESP8266 board to publish on Facebook. The status update that was used in this recipe is `Status update from ESP8266 via Temboo #IoT`.

15. Click on the **Generate** code button to test if the Choreo successfully publishes the status update. If the status update is published successfully, you will receive a positive response in the output section with the ID of the transaction. Better still, you can check your Facebook account to see if the status update was posted.

16. Download the `sketch` and `header` file. Also, remember to update the sketch so that it is compatible with your ESP8266. The `sketch` is as follows.

▶   Include the `ESP8266` and `Temboo` libraries, the `header` file, and other supporting libraries:

```
#include <SPI.h>
#include <ESP8266WiFi.h>
#include <WiFiClient.h>
#include <Temboo.h>
#include "TembooAccount.h" // Contains Temboo account information
```

- Create a `WiFiClient` object:

```
WiFiClient client;
```

- Initialize the number of calls and the maximum number of calls to be made:

```
int calls = 1;    // Execution count, so this doesn't run forever
int maxCalls = 10;    // Maximum number of times the Choreo should
be executed
```

- Initialize a serial interface and connect to the Wi-Fi network:

```
void setup() {
  Serial.begin(115200);
  Serial.print("Connecting to Wi-Fi hotspot");

  WiFi.begin(WIFI_SSID, WPA_PASSWORD);
  while (WiFi.status() != WL_CONNECTED) {
    delay(500);
    Serial.print(".");
  }
  Serial.println("");
  Serial.println("WiFi connected");
  Serial.println("IP address: ");
  Serial.println(WiFi.localIP());
  delay(1000);
}
```

- Post a status update every 30 seconds a maximum of 10 times:

```
void loop() {
  if (calls <= maxCalls) {
    Serial.println("Running SetStatus - Run #" + String(calls++));
```

- Create a `TembooChoreo` object and configure it:

```
    TembooChoreo SetStatusChoreo(client);

    // Invoke the Temboo client
    SetStatusChoreo.begin();

    // Set Temboo account credentials
    SetStatusChoreo.setAccountName(TEMBOO_ACCOUNT);
    SetStatusChoreo.setAppKeyName(TEMBOO_APP_KEY_NAME);
    SetStatusChoreo.setAppKey(TEMBOO_APP_KEY);
    SetStatusChoreo.setDeviceType(TEMBOO_DEVICE_TYPE);
```

- ▶ Set the Choreo inputs:

```
// Set Choreo inputs
String MessageValue = "Status update from ESP8266 via Temboo
#IoT";
SetStatusChoreo.addInput("Message", MessageValue);
String AccessTokenValue = "EAAF7Arxxxxxxxxxxxxxxxxxxxxx";
SetStatusChoreo.addInput("AccessToken", AccessTokenValue);
```

- ▶ Select the `Choreo` that is going to be run. In this case, it is the `SetStatus` `Choreo` for Facebook:

```
// Identify the Choreo to run
SetStatusChoreo.setChoreo("/Library/Facebook/Publishing/
SetStatus");
```

- ▶ Run the Choreo and listen for any incoming data from the Facebook server:

```
// Run the Choreo; when results are available, print them to
serial
SetStatusChoreo.run();

while(SetStatusChoreo.available()) {
  char c = SetStatusChoreo.read();
  Serial.print(c);
}
SetStatusChoreo.close();
}

Serial.println("\nWaiting...\n");
delay(30000); // wait 30 seconds between SetStatus calls
}
```

- ▶ Header file:

```
#define TEMBOO_ACCOUNT "yhtomit"  // Your Temboo account name
#define TEMBOO_APP_KEY_NAME "myFirstApp"  // Your Temboo app key
name
#define TEMBOO_APP_KEY "albLWxxxxxxxxxxxxxxxxxxxxxxxxx"  // Your
Temboo app key
#define TEMBOO_DEVICE_TYPE "a101+w101"

#define WIFI_SSID "ssid"
#define WPA_PASSWORD "password"

#if TEMBOO_LIBRARY_VERSION < 2
#error "Your Temboo library is not up to date. You can update it
using the Arduino library manager under Sketch > Include Library >
Manage Libraries..."
#endif
```

17. The `Temboo` library that comes with the Arduino IDE does not support ESP8266 boards, and will probably bring up compiling errors when you try to upload it. To solve this, download an updated version of the `Temboo` library that works with ESP8266 from `https://github.com/marcoschwartz/esp8266-IoT-cookbook`.

18. Copy the sketch to your Arduino IDE and change the SSID in the `header` file from `ssid` to the name of your Wi-Fi network and the password from `password` to the password of your Wi-Fi network.

19. Upload the sketch to your ESP8266 board.

20. Open the serial monitor so that you can view the incoming data.

How it works...

The program connects to the Wi-Fi network and posts a status update to Facebook every 30 seconds. This is made possible by the `SetStatus Choreo` for Facebook. When the Choreo is called in the program, the ESP8266 connects to Temboo and, using the `AccessToken`, publishes data to Facebook.

See also

If you want to log a lot of data online for future reference, you will need a better solution than using social media platforms. If that is the case, you may consider using Google Drive for your data storage needs. The next recipe will look at how to go about *Storing data on Google Drive*.

Storing data on Google Drive

In this recipe, we will look at how to store data on Google Drive directly from an ESP8266 board. This usually comes in handy when you need to log data continuously, over a period of time, from your IoT project. The data can be stored in different formats on Google Drive. So you can store virtually any kind of data, ranging from digital sensor inputs to images. Here we will look at how to log analog sensor input to a spreadsheet on Google Drive.

Getting ready

You will need the following hardware components:

- ESP8266 board
- USB cable
- 1 kΩ-10 kΩ photocell (`https://www.sparkfun.com/products/9088`)
- 1 kΩ resistor
- Breadboard
- Jumper wires

The setup will resemble the one in *Tweeting data from the Arduino board* in this chapter, where you used your ESP8266 board to read analog input from a **Light Dependent Resistor** (**LDR**) connected to a 10 kΩ resistor in a voltage divider circuit.

Also, you need to create a spreadsheet on Google Drive. To do that, visit `https://docs.google.com/spreadsheets/u/0/`. Under the **Start a new spreadsheet** section on the top of the page, click on the **Blank** icon. This will open a new blank spreadsheet. Enter the name of the spreadsheet in the area provided on the top of the page. The name of the spreadsheet in this recipe is `ESP8266_Log`. You can also enter the column headings, though this is optional. The column headings in this recipe are `Index` and `Value`:

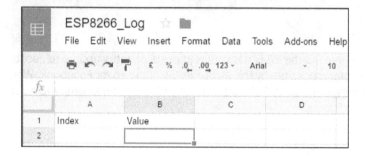

You need a Google account to create a spreadsheet on Google Drive. So, if you do not have one, you can sign up here:

`https://accounts.google.com/NewAccount`.

How to do it...

Now that you have your hardware and Google spreadsheet ready, you can start working on the code. You will use Temboo Choreos to create the `sketch` and `header` files:

1. So, start by logging on to the Temboo platform and selecting **Google | SpreadSheets** from the `libraries` menu.

2. You will then log in to Google developer console using your Google account and create a new project. Access the **API manager** and, under the **Library** tab, select **Google Drive API** (or **Drive API** in the **Google Apps APIs** section), then enable it.

3. Click on the **Credentials** tab, create a new **Client ID**, and specify that it is a **Web application**. Provide the name, and under **Authorised redirect** URIs, enter `https://{Your Temboo Account Name}.temboolive.com/callback/google`. Then click on **create** to complete the process.

4. Select the **OAuth consent screen** under the **Credentials** tab. Enter your e-mail address and the product name. The rest of the fields are optional. Then click on **Save**. Now you have finished setting up your Google project:

The name of the **Client ID** in this recipe was `ESP8266_Log` and the product name was `ESP8266`. You can use whichever name you like:

1. Having obtained the **Client ID**, you can open the InitializeOAuth Choreo on Temboo by selecting **Google | OAuth | InitializeOAuth** from the libraries menu.

2. Enter the **Client ID** from the Google project and the **scope**. The scope in this case is `https://spreadsheets.google.com/feeds/`.

3. Click on **Generate Code**. This is going to output the AuthorizationURL and the **CallbackID**. Open the AuthorizationURL in a different webpage and grant your application access to your Google account.

4. Access the FinalizeOAuth Choreo from **Google | OAuth| FinalizeOAuth** in the libraries menu.

5. Input the **Client ID** and **Client Secret** from the Google application and enter the **CallbackID** that was generated by the InitializeOAuth Choreo.

6. Click on the **Generate code** button to generate the `AccessToken`.

7. Select **Google | Spreadsheets | AppendRow** from the libraries menu. This opens the `AppendRowChoreo` in which you will input `Client ID`, `Client Secret`, `RefreshToken (AccessToken)`, `RowData` and `SpreadsheetTitle`.

8. You can test if the Choreo is working, by clicking on the **Generate Code** button. If data is posted successfully on your spreadsheet, then you can define the device you are using and the method of connection so that the Arduino sketch can be generated.

In this recipe, the board that was selected was the **Arduino 101** and the connection was the Arduino Wi-Fi 101 shield, which we named ESP8266. You can refer to *Tweeting data from the Arduino board* to see how that was done.

Download the source code and edit it so that it is compatible with the ESP8266 board. The edited sketch is shown here:

▸ Include `ESP8266` libraries and create a `WiFiClient` object:

```
#include <SPI.h>
#include <ESP8266WiFi.h>
#include <WiFiClient.h>
#include <Temboo.h>
#include "TembooAccount.h" // Contains Temboo account information

WiFiClient client;
```

▸ Declare variables:

```
int calls = 1;    // Execution count, so this doesn't run forever
int maxCalls = 10;    // Maximum number of times the Choreo should
be executed
int sensorVal = 0; // holds sensor analog reading
Initialize serial communication and connect ESP8266 to the Wi-Fi
hotspot
void setup() {
  Serial.begin(115200);
  Serial.print("Connecting to Wi-Fi hotspot");

  WiFi.begin(WIFI_SSID, WPA_PASSWORD);
  while (WiFi.status() != WL_CONNECTED) {
    delay(500);
    Serial.print(".");
  }
  Serial.println("");
  Serial.println("WiFi connected");
  Serial.println("IP address: ");
  Serial.println(WiFi.localIP());
  delay(1000);
}
```

▸ Read sensor input and post it to Google spreadsheets after every 30 seconds, a maximum of 10 times:

```
void loop() {
  if (calls <= maxCalls) {
    Serial.println("Running AppendRow - Run #" + String(calls++));
    sensorVal = analogRead(A0); // get sensor reading
    TembooChoreoSSL AppendRowChoreo(client);

    // Invoke the Temboo client
```

```
    AppendRowChoreo.begin();

    // Set Temboo account credentials
    AppendRowChoreo.setAccountName(TEMBOO_ACCOUNT);
    AppendRowChoreo.setAppKeyName(TEMBOO_APP_KEY_NAME);
    AppendRowChoreo.setAppKey(TEMBOO_APP_KEY);
    AppendRowChoreo.setDeviceType(TEMBOO_DEVICE_TYPE);

Set the Choreo inputs and run the Choreo
    // Set Choreo inputs
    String SpreadsheetTitleValue = "ESP8266_Log";
    AppendRowChoreo.addInput("SpreadsheetTitle",
SpreadsheetTitleValue);
    // generate row data
                    // Index column = number of calls
                    // value column = sensor reading
    String RowDataValue = String(calls)+","+String(sensorVal);
    AppendRowChoreo.addInput("RowData", RowDataValue);
    String RefreshTokenValue = "ya29.xxxxxxxxxxxxxxx";
    AppendRowChoreo.addInput("RefreshToken", RefreshTokenValue);
    String ClientSecretValue = "oWHEGxxxxxxxxxxxxxxx";
    AppendRowChoreo.addInput("ClientSecret", ClientSecretValue);
    String ClientIDValue = "974405649595-xxxxxxxxxxxx ";
    AppendRowChoreo.addInput("ClientID", ClientIDValue);

    // Identify the Choreo to run
    AppendRowChoreo.setChoreo("/Library/Google/Spreadsheets/
AppendRow");

    // Run the Choreo; when results are available, print them to
serial
    AppendRowChoreo.run();

Print any data that is received from the Google Drive servers
    while(AppendRowChoreo.available()) {
      char c = AppendRowChoreo.read();
      Serial.print(c);
    }
    AppendRowChoreo.close();
  }

  Serial.println("\nWaiting...\n");
  delay(30000); // wait 30 seconds between AppendRow calls
}
```

▶ Header file:

```
#define TEMBOO_ACCOUNT "yhtomit"  // Your Temboo account name
#define TEMBOO_APP_KEY_NAME "myFirstApp"  // Your Temboo app key
name
#define TEMBOO_APP_KEY "albLWVSvxxxxxxxxxxxxxxxxxxx" // Your Temboo
app key
#define TEMBOO_DEVICE_TYPE "a101+w101"

#define WIFI_SSID "ssid"
#define WPA_PASSWORD "pass"

#if TEMBOO_LIBRARY_VERSION < 2
#error "Your Temboo library is not up to date. You can update it
using the Arduino library manager under Sketch > Include Library >
Manage Libraries..."
#endif
```

The `Temboo` library that comes with the Arduino IDE does not support ESP8266 boards, and will probably bring up compiling errors when you try to upload it. To solve this, download an updated version of the `Temboo` library that works with ESP8266 from `https://github.com/marcoschwartz/esp8266-IoT-cookbook`.

1. Copy the sketch to your Arduino IDE and change the SSID in the `header` file from `ssid` to the name of your Wi-Fi network and the password from `password` to the password of your Wi-Fi network.

2. Upload the sketch to your ESP8266 board. Open the serial monitor so that you can view the incoming data.

The following screenshot shows the spreadsheet after the ESP8266 posts the first sensor reading:

How it works...

The sketch includes the `ESP8266` and `Temboo` libraries and connects the ESP8266 board to the Wi-Fi network whose credentials you have provided in the `header` file. The program then reads the sensor input on the analog pin in 30-second intervals, and then posts the read input to the spreadsheet on Google Drive, using the `AppendRowChoreo`. This process is repeated 10 times.

See also

So far we have been looking at Temboo and how to use Choreos to post data to online platforms. However, there are also other IoT platforms that we can use to achieve that. IFTTT is a good example. We will explore it in the next recipe.

Automation with IFTTT

If This Then That (**IFTTT**) is an online service that allows users to create simple conditional statements called applets. The applets are triggered depending on changes made on or to other web services. For instance, you can set an IFTTT applet that sends you an e-mail if a user includes a certain hashtag in their tweets. You can also configure an applet that saves all the photos that you have been tagged in on Facebook to a cloud server such as Google Drive.

In this recipe, we will look at how to use the IFTTT online service to automate IoT projects. We will use it to publish data to different online platforms, just like Temboo. To demonstrate how to do that, we are going to publish temperature data from a DHT11 sensor to Twitter, using an ESP8266 board.

Getting ready

You will need the following hardware components:

- ESP8266 board
- USB cable
- DHT11 temperature/humidity sensor (`https://www.sparkfun.com/products/10167`)
- 10 kΩ resistor
- Breadboard
- Jumper wires

The DHT11 pin configuration is shown in the following figure:

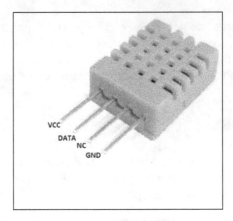

1. First, mount the ESP8266 board and the DHT11 sensor onto the breadboard.

2. Connect a 10 kΩ pull up resistor to the DHT11 data pin and connect the VCC pin and **GND** pin to the 3V pin and **GND** pin of the ESP8266 board, respectively.

3. Finally, connect the data pin of the DHT11 to GPIO 2 of the ESP8266 board. Use jumper wires for the connections.

The setup is shown in the following diagram:

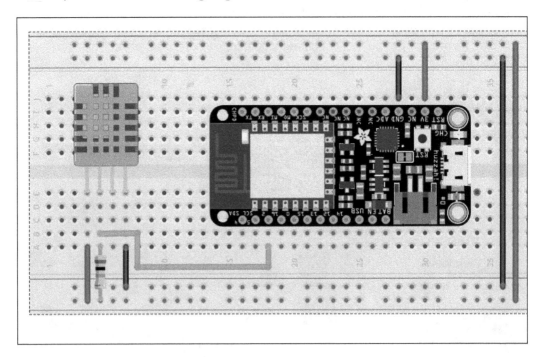

Now that the hardware setup is complete, you should configure IFTTT. Start by creating an account on IFTTT. To do so, go to the official website, `https://ifttt.com`, click on the **sign up** button, and complete the sign up form by entering your e-mail address and password.

How to do it...

Once you have an IFTTT account and are logged in follow these steps:

1. Click on the **drop-down** menu on the top-right corner of the page and select **New Applet**. Click on **this** on the page that appears:

This should bring up the create page that provides you with a list of all the services/ channels that are supported by IFTTT. We will be using the maker channel, so just search for `maker` in the services page. This should bring up a couple of choices. Select the **Maker** channel that has this icon:

2. Click on the **Connect** button on the page that appears. This leads you to a page that prompts you to create a trigger that fires every time **Maker** service gets a web request.

3. Click on **Receive a web request** and specify the event name which will be sending a web request to **Maker** service. We will call ours `temperature_low`.

4. Once you have entered the name, click on the **Create trigger** button.

5. Now proceed to create the action to be taken when the trigger is fired. To do that, click on the **that** link:

Select the **Twitter** service and select the **post a tweet** action. Provide the tweet that will be posted when the trigger is fired. It should respect the Twitter guidelines for posting tweets. Most important of all is not to exceed 140 characters. The tweet we created is `{{Value1}} degree Celsius ==> Temperature Low! #ESP8266 #IoT #IFTTT`.

1. You can display a value from your sensor on the tweet, by entering it as `{{Value1}}`. The number of values may differ depending on the parameters from the ESP8266 board that you would like to post on Twitter. In our case we will only have one value, which is the temperature in degrees Celsius.

2. Click on **Create action** and, if you are satisfied with the settings, click on **Finish** to create the applet.

3. Go to `https://ifttt.com/maker` to get the secret key:

    Account Info

    Connected as: yhtomit

    URL: https://maker.ifttt.com/use/

    Status: active

    Edit connection

    ---

4. The key is in the provided URL. Open that URL in a different tab on your web browser to see how to use the secret key to fire a trigger. You will fire the trigger using this URL format: `https://maker.ifttt.com/trigger/{event}/with/key/{your key}`.

5. If you are going to be passing some values with your trigger URL, as in our case, the format will be as follows: `https://maker.ifttt.com/trigger/{event}/with/key/{your key}?value1={val}`.

 Replace `{event}` with the name of your event as you called when setting up your Maker channel, `{your key}` with your unique key and `{val}` with the value of whatever parameter you are monitoring (in our case, it's temperature). You can run the URL on a web browser and check your Twitter to see if the tweet appears on your timeline.

The code will be as follows:

1. Include the `ESP8266` library and the `DHT` library:

```
#include <ESP8266WiFi.h>
#include "DHT.h"
Set time between each trigger. Currently set to 30 seconds:
#define timeInterval 30000 // time between each trigger
```

2. Define the temperature sensor signal pin and sensor type, and create an object of the `DHT` library:

```
#define DHTPIN 2       // what digital pin we're connected to
#define DHTTYPE DHT11   // DHT 11
DHT dht(DHTPIN, DHTTYPE);
Set Wi-Fi network credentials:
const char* ssid     = "ssid";
const char* password = "pass";
Set IFTTT requirements:
const char* host = "maker.ifttt.com";
const char* privateKey = "bqT9lyXcojqIexd8DiYk1F";
const char* event = "temperature_low";
```

3. Variable to hold previous time `trigger` was fired:

```
long lastTime = 0; // holds previous time trigger was sent
Initialize serial communication and the dht sensor and connect
ESP8266 module to the Wi-Fi network:
void setup() {
  Serial.begin(115200); // initialize serial communication
  dht.begin(); // initialize DHT11 sensor
  delay(100);

  // We start by connecting to a WiFi network

  Serial.println();
  Serial.println();
  Serial.print("Connecting to ");
  Serial.println(ssid);

  WiFi.begin(ssid, password);

  while (WiFi.status() != WL_CONNECTED) {
    delay(500);
    Serial.print(".");
  }
```

```
      Serial.println("");
      Serial.println("WiFi connected");
      Serial.println("IP address: ");
      Serial.println(WiFi.localIP());
  }
```

4. Get the `temperature` measurement from the `dht` sensor:

```
void loop() {
  // Read sensor inputs
  // get temperature reading in Celsius
  float temperature = dht.readTemperature();
  // Check if any reads failed and exit early (to try again).
  while (isnan(temperature)) {
    Serial.println("Failed to read from DHT sensor!");
    delay(2000); // delay before next measurements
    //get the measurements once more
    temperature = dht.readTemperature();
  }
```

5. Check if the `temperature` has gone down by 30 degrees Celsius and if the last time the trigger was fired was 30 seconds ago. If those conditions are both true, the trigger can be fired:

```
if(temperature < 30 && millis()- lastTime > timeInterval){ //
temperature is less than 30 deg celsius
    Serial.print("connecting to ");
    Serial.println(host);
```

6. Connect to the `host serve:r`:

```
      // Use WiFiClient class to create TCP connections
      WiFiClient client;
      const int httpPort = 80;
      if (!client.connect(host, httpPort)) {
        Serial.println("connection failed");
        return;
      }
```

7. Create the URL that will be used to fire the `trigger`:

```
// We now create a URI for the request
String url = "/trigger/";
url += event;
url += "/with/key/";
url += privateKey;
url += "?value1=";
url += String(temperature);

Serial.print("Requesting URL: ");
Serial.println(url);
```

8. Send `HTTP` request to fire the `trigger` and read the incoming response from the server:

```
// This will send the request to the server
client.print(String("GET ") + url + " HTTP/1.1\r\n" +
             "Host: " + host + "\r\n" +
             "Connection: close\r\n\r\n");
unsigned long timeout = millis();
while (client.available() == 0) {
  if (millis() - timeout > 5000) {
    Serial.println(">>> Client Timeout !");
    client.stop();
    return;
  }
}

// Read all the lines of the reply from server and print them
to Serial
while(client.available()){
  String line = client.readStringUntil('\r');
  Serial.print(line);
}
Serial.println();
Serial.println("closing connection");
lastTime = millis(); // save time of last trigger
  }
}
```

9. Copy the sketch to your Arduino IDE and change the SSID in the sketch from `ssid` to the name of your Wi-Fi network and the password from `password` to the password of your Wi-Fi network.

10. Upload the sketch to your ESP8266 board. Open the serial monitor so that you can view incoming data.

How it works...

The program connects the ESP8266 module to the Wi-Fi network, then temperature data is obtained from the DHT11 sensor. If the temperature is less than 30 degrees Celsius and the last trigger was sent more than 30 seconds ago, the program connects to the IFTTT server and sends an HTTP request that fires the trigger.

When the trigger is fired, the IFTTT posts a tweet on your Twitter account containing the information you provided when setting up the Twitter channel on the IFTTT applet.

There's more...

Obtain more measurements on your project, such as temperature in Fahrenheit and humidity, then pass those values to your tweet on IFTTT. This way, when the trigger is fired, these measurements are also displayed on your tweets.

See also

The ESP8266 can also be used to send push notifications to your phone through IFTTT. If you are keen to achieve that, proceed to the next recipe for guidelines on how to do so.

Sending push notifications

Using IFTTT, you can easily send push notifications to your phone from your ESP8266-based IoT project. This is possible through the Pushover service. To demonstrate this, we will get a temperature measurement from the DHT11 sensor using the ESP8266 and send push notifications to an Android phone when the temperature goes below 30 degrees Celsius.

Getting ready

You will use the same setup as in the previous recipe, which includes a DHT11 sensor connected to an ESP8266 board. Also, download the Pushover app on your phone, create an account, and provide the name of your device.

How to do it...

1. Log in to your IFTTT account and create a new applet.

2. Set the Maker service as your This, as you did in the previous recipe. The event name will still be `temperature_low`. The Maker service configuration will be as follows:

3. Once the Maker service is fully configured, proceed to set up the That service. The service we will be using to send the push notifications is the Pushover service.

4. So, click on **That**, then search for the Pushover service in the services page and select it. If you haven't already logged on to Pushover, you will be required to do so, and also select the device your Pushover application is installed in. Once that is done, you can proceed with setting up the Pushover service on IFTTT.

5. Having logged on to the Pushover service successfully, click on **Send a Pushover notification**.

6. Enter the required settings for your Pushover message. Some of the information you are supposed to provide is the title, the message, message priority, the sound of the push notification, and the device the push notification will be sent to. In this recipe, the settings used were as follows:

   - Title: `{{EventName}}`
   - Message: `{{Value1}}` degrees Celsius
   - Message priority: Normal
   - Message sound: Pushover(default)
   - Device: SonyXperia-Z

Entering `{{EventName}}` and `{{Value1}}` in the textboxes automatically sets the event name that you provided and the value that is going to be in the trigger URL.

1. Once you finish configuring the Pushover service, click on the **Create action** button, then click on **Finish**. Go to `https://ifttt.com/services/maker/settings` to get the secret key. Use the secret key to create a `trigger` HTTP request such as `https://maker.ifttt.com/trigger/{event}/with/key/{your_key}?value1={temperature_measurement}`.

   We will use the same code as in the previous recipe. However, this time the `trigger HTTP` request will send push notifications to your phone when the temperature goes down by 30 degrees Celsius.

   The push notifications will appear as follows:

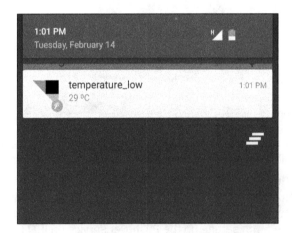

2. When you touch it, it will open the Pushover app and show the message shown in the following screenshot:

How it works...

The program connects the ESP8266 module to the Wi-Fi network, then temperature data is obtained from the DHT11 sensor. If the temperature is below 30 degrees Celsius and the last trigger was sent more than 30 seconds ago, the program connects to the IFTTT server and sends an HTTP request that fires the trigger.

When the trigger is fired, the IFTTT sends a push notification to your phone containing the information you provided when setting up the Pushover channel on the IFTTT applet.

There's more...

You can spice up your project by adding two more sensor values to be displayed on the push notification on your phone. Moreover, you can change the code so that you receive push notifications when the temperature goes over 30 degrees Celsius.

See also

When using IFTTT, you can send e-mail notifications to let you know when a sensor value has exceeded or gone below a certain threshold. Proceed to the next recipe to learn how to do that.

Sending e-mail notifications

In this recipe, you will learn how to send e-mail notifications from your ESP8266 board. To achieve that, we will use the e-mail service on the IFTTT platform. This way, we will receive an e-mail from the ESP8266 whenever the temperature goes down by 30 degrees Celsius.

Getting ready

You will need the following hardware components:

▸ ESP8266 board

▸ USB cable

▸ DHT11 temperature/humidity sensor (`https://www.sparkfun.com/products/10167`)

▸ 10 kΩ resistor

▸ Breadboard

▸ Jumper wires

The setup will be the same as in *Automation with IFTTT* of this chapter:

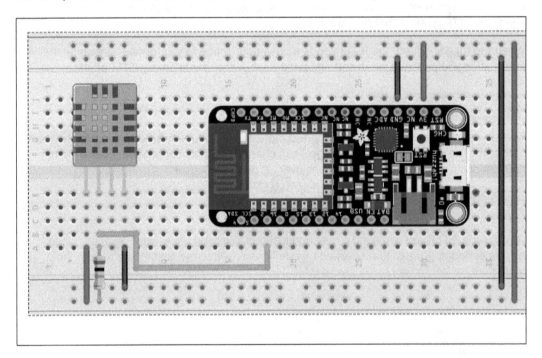

If you do not have an e-mail account, start by creating one on the e-mail service provider of your choice. In this recipe, we will be using a Gmail account.

How to do it...

1. Create a new applet on IFTTT and click on **this** to select the service that is going to be monitored. Select the **Maker** service and configure it (refer back to *Automation with IFTTT*).

2. Once the setup is complete, click on **that** to select the service whose actions are going to be triggered. In this recipe, it will be the e-mail service. Therefore, search for it on the services page and select it.

3. Choose the **Send me an email** action of the e-mail service and set the **Subject** of the e-mail and the **Body**:

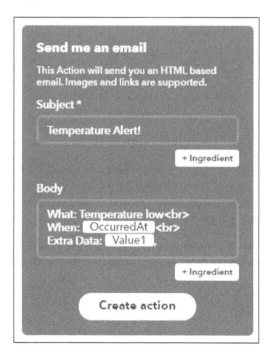

4. Once you are satisfied with your configuration, click on the **Create action** button. You can then review your app, and if you are satisfied with it, click on the **Finish** button.

5. Visit `https://ifttt.com/services/maker/settings` to get the secret key that you will use to create the trigger URL. Our trigger URL is `https://maker.ifttt.com/trigger/{event}/with/key/{your_key}?value1={temperature measurement}`.

The event name in this recipe will still be `temperature_low`, and the sketch will be similar to the one in *Automation with IFTTT*. However, in this case, when the temperature goes down by 30 degrees Celsius, the ESP8266 board will send you an e-mail notification, as shown in the following screenshot:

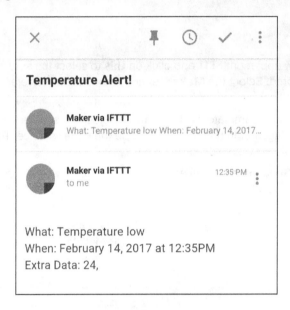

How it works...

The program connects the ESP8266 module to the Wi-Fi network, then temperature data is obtained from the DHT11 sensor. If the temperature is down by 30 degrees Celsius and the last trigger was sent more than 30 seconds ago, the program connects to the IFTTT server and sends an HTTP request that fires the trigger.

When the trigger is fired, the IFTTT sends an e-mail notification to your phone containing the information you provided when setting up the e-mail service on the IFTTT applet.

See also

The ESP8266 can also be used to send text message notifications. The next recipe will show you how to do that using the IFTTT platform.

Sending text message notifications

Getting text message notifications is probably the most reliable way to deliver real-time data and information. This is because the phone doesn't need an Internet connection to receive the notifications. With IFTTT, you can use your ESP8266 board to send text messages from your IoT projects to your phone. To demonstrate this, we will send text message notifications to a phone from the ESP8266, when the temperature measurement from the DHT11 sensor goes below 30 degrees Celsius.

Getting ready

You will use a hardware setup similar to the one in _Automation with IFTTT_. Also, you will have to set up the SMS service on IFTTT.

How to do it...

1. Create a new applet on IFTTT and click on **this** to select the service that is going to be monitored. Select the **Maker** service and configure it (refer back to _Automation with IFTTT_).

 Remember to call the event `temperature_low`. If you change the event name, you have to also change it in the Arduino code.

2. Once the setup is complete, click on **that** to select the service whose actions are going to be triggered. In this recipe, we will be using the SMS service.

3. If it is your first time connecting to the SMS service, you will be required to provide your phone number and then click on **Send pin**:

4. A 4-digit pin will be sent to your phone. Enter the 4-digit pin in the text box provided, then click on **Connect**. This will take you to the **Choose action** page where you will click on the **Send me an SMS** option. You will then write the message that you would like to receive on your phone:

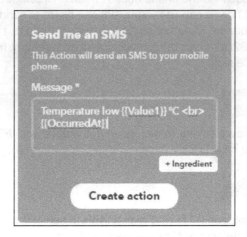

5. When you finish composing the message, click on the **Create action** button. You can then review your app, and if you are satisfied with it, click on the **Finish** button.

6. Visit `https://ifttt.com/services/maker/settings` to get the secret key that you will use to create the trigger URL. Our trigger URL is `https://maker.ifttt.com/trigger/{event}/with/key/{your_key}?value1={temperature measurement}`.

The sketch will be similar to the one in *Automation with IFTTT*. When the temperature goes down by 30 degrees Celsius, the ESP8266 board sends a text message notification to your phone, as shown in the following screenshot:

How it works...

The program connects the ESP8266 module to the Wi-Fi network, then temperature data is obtained from the DHT11 sensor. If the temperature is below 30 degrees Celsius and the last trigger was sent more than 30 seconds ago, the program connects to the IFTTT server and sends an HTTP request that fires the `trigger`.

When the `trigger` is fired, the IFTTT sends a text message notification to your phone containing the information you provided when setting up the SMS service on the IFTTT applet.

There's more...

Get the humidity measurement from the DHT11 sensor and append it to the text message.

See also

There are some common problems that you may face when using web services. The next recipe will describe them and show you how to solve them.

Troubleshooting common issues with web services

In this recipe, we will discuss some of the problems you may run into, and how to troubleshoot and solve them.

The board is not connecting to the Wi-Fi network

This usually happens if the Wi-Fi SSID and password provided in the code do not match those of the Wi-Fi network the ESP8266 is supposed to connect to. You can solve this by providing the correct credentials in your code.

The generate code button on Temboo returns an error when pressed the second time

Choreos, especially those of social media platforms, may return an error when they are tested more than once. This error is returned by social media platform servers because they have detected a duplicate update. To solve the issue, change the data that is being updated to the social media platform before clicking on the generate code button again.

The Temboo sketches bring up errors when I try to compile them

This happens because the `Temboo` library that comes with the Arduino IDE does not support ESP8266 boards. To solve this issue, download an updated version of the `Temboo` library from `https://github.com/marcoschwartz/esp8266-IoT-cookbook`. Replace the original `Temboo` library with it.

All my applets on IFTTT are being activated when the trigger is fired

This is caused by using duplicate event names on your applets. To solve the issue, change the event name of your applet.

The IFTTT SMS service is not working

This can be caused by two things. Either the mobile phone carrier you are using is not supported by the SMS service, or you have exceeded the limit for the number of SMSes you can send in a month. The limit is 100 SMSes per month for USA and Canadian residents and 10 SMSes per month for people living outside North America.

7

Machine to Machine Interactions

In this chapter, we will cover:

- ▸ Types of IoT interactions
- ▸ Basic local machine to machine interaction
- ▸ Cloud machine to machine interaction
- ▸ Automated M2M with IFTTT
- ▸ M2M alarm system
- ▸ Automated light control
- ▸ Automated gardening controller
- ▸ Troubleshooting common issues with web services

Introduction

Apart from storage of data and sending alerts, the IoT can be used to build elaborate systems that work autonomously without human monitoring or intervention. This is possible through **Machine to Machine** (**M2M**) interactions, where different IoT devices communicate with each other either directly or indirectly, and carry out actions depending on the information/data they receive. This chapter is going to look at M2M interactions in detail and how they can be applied in IoT projects.

Types of IoT interactions

The Internet of Things (IoT) is a vast network of devices that share information using different Internet protocols. It is comprised of different kinds of interactions between devices, online servers, and humans. Most interactions involve communication between one IoT device and another, or communication between an IoT device and a human. Both forms of interaction can either be direct or via a web server (indirect).

Though human interaction in IoT systems is important, it is not always needed. You can create autonomous systems that do not require any human input to operate. This is made possible by M2M interaction. In such systems, the different IoT devices send data to each other and execute different actions depending on the input they receive. This eliminates the need for human intervention.

M2M is a term that is used to describe technology that allows communication between networked devices. The devices use network resources to exchange data with remote application infrastructure for monitoring and control purposes. M2M works independently, without the need for human control, and is the basis for the concept known as the Internet of Things. In fact, IoT can be loosely described as machine to machine communication via Internet protocols.

The two main types of machine to machine interactions are:

- Local machine to machine interaction
- Cloud machine to machine interaction

Local machine to machine interaction involves the direct exchange of data between devices in the same locality. Both devices must connect to the same wireless network for communication to take place. This network can either be the local Wi-Fi hotspot or an ad Hoc network that is hosted by one of the devices.

In local M2M interaction, one device is configured as a web server. The other devices are configured as clients. The web server manages the communication between the clients in the network. This kind of interaction does not require an Internet connection.

Cloud machine to machine interaction is used for communication between devices located in remote areas, provided there is an active Internet connection. Unlike local M2M, devices do not interact directly with each other. A cloud server acts as an intermediary between the communicating devices.

In cloud M2M, the communicating devices are configured as clients. One device sends data to the cloud server, then the other device reads the data from the cloud server. That way devices are able to share data and information even when they are thousands of miles apart.

One protocol that facilitates cloud M2M is MQTT. It allows devices to publish data using a unique topic to a cloud server known as an MQTT broker. The receiving devices then subscribe to the topic where the data was published and receive the latest published data.

To learn more about machine to machine interactions, proceed to the next recipe for guidelines on how to implement *Basic local machine to machine interactions*.

Basic local machine to machine interactions

In this recipe, we are going to demonstrate *Basic local machine to machine interactions*. To do that, we will build a simple project that demonstrates machine to machine interaction between two ESP8266 boards. This will enable you to create simple networks using your ESP8266 boards, with no Internet connection.

Getting ready

You will need the following hardware components for this project:

- Two ESP8266 boards
- Two USB cables
- 220 Ω resistor
- LED
- Momentary **Push** button
- 10 kΩ resistor

Connect a **Push** button and a 10 kΩ pull up resistor to the GPIO2 pin. The **Push** button will be used as an input. This ESP8266 board will be configured as the client. The setup is shown in the following figure:

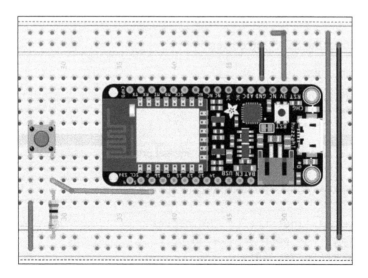

For the second ESP8266, connect an LED to the GPIO2 pin via a 220 Ω current limiting resistor. This ESP8266 is going to be configured as the server. The server setup will look like this:

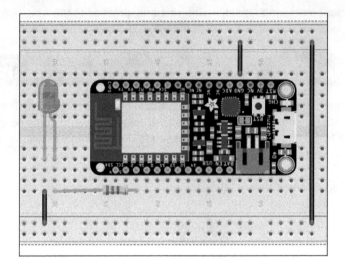

How to do it...

To successfully transfer data between the two ESP8266 boards, set up one of the boards as a client and the other one as the server. You will configure the client to send a request to the server. The request will correspond to the state of the input pin GPIO2 of the ESP8266.

Configure the other ESP8266 as a webserver and use it to create an access point so that the client connects directly to it. This will eliminate the need for an external Wi-Fi hotspot. The ESP8266 that is configured as a server will change the status of the output pin GPIO2 to correspond to the request received from the client.A reply will be sent to the clientstating the current status of the output pin GPIO2 of the ESP8266.

Here is the sketch for the ESP8266 server:

1. Include libraries:

```
#include <ESP8266WiFi.h>
#include <WiFiClient.h>
#include <ESP8266WebServer.h>
```

2. Set the ssid and password for you access point:

```
const char* ssid = "hotspot_ssid";
const char* password = "hotspot_password";

// Create an instance of the server
// specify the port to listen on as an argument
WiFiServer server(80);
```

3. Initialize the serial communication port:

```
void setup() {
  delay(1000);
  Serial.begin(115200);
  delay(10);
```

▶ Set GPIO2 as output:

```
// prepare GPIO2
pinMode(2, OUTPUT);
digitalWrite(2, 0);
```

▶ Start the access point using the SSID and `password` that you provided earlier:

```
Serial.print("Configuring access point...");
/* You can remove the password parameter if you want the AP to
be open. */
WiFi.softAP(ssid, password);

IPAddress myIP = WiFi.softAPIP();
Serial.print("AP IP address: ");
Serial.println(myIP);

server.begin();
Serial.println("HTTP server started");
}
```

▶ Check the `client` has connected to the `server` and read any incoming data from the `client`:

```
void loop() {
  // Check if a client has connected
  WiFiClient client = server.available();
  if (!client) {
    return;
  }

  // Wait until the client sends some data
  Serial.println("new client");
  while(!client.available()){
    Serial.print('.');
    delay(1);
  }

  // Read the first line of the request
  String req = client.readStringUntil('\r');
  Serial.println(req);
  client.flush();
```

- ▸ Evaluate the `request` from the client to determine what state the GPIO2 pin should be set at:

```
// Match the request
int val;
if (req.indexOf("/gpio/0") != -1)
  val = 0;
else if (req.indexOf("/gpio/1") != -1)
  val = 1;
else {
  Serial.println("invalid request");
  client.stop();
  return;
}
```

- ▸ Set the `GPIO2` pin state to correspond to the request from the `client`:

```
// Set GPIO2 according to the request
digitalWrite(2, val);

client.flush();
```

- ▸ Generate the `response` that the server will send to the client:

```
// Prepare the response
String s = (val)?"high":"low";
```

- ▸ Send the response to the `client` and end the session:

```
// Send the response to the client
client.print(s);
delay(1);
Serial.println("Client disonnected");

// The client will actually be disconnected
// when the function returns and 'client' object is detroyed
}
```

Upload this sketch to the ESP8266 board that will be used as a server. Remember to change `hotspot_ssid` and `hotspot_password` to your preferred access point SSID and password respectively, before uploading the code. Open the serial monitor and note down the IP address of your ESP8266 server. You will include it in the ESP8266 client sketch.

The sketch for your ESP8266 client is as follows:

- ▸ Include library:

```
#include <ESP8266WiFi.h>
```

- ▸ Set SSID and `password` for your server's access point:

```
const char* ssid     = "your-ssid";
const char* password = "your-password";
```

- ▸ Set the IP address of your ESP8266 server:

```
const char* host = "192.XXX.XXX.XXX";
```

- ▸ Initialize the serial communication port, set `GPIO2` as an input, and connect to the Wi-Fi access point:

```
void setup() {
  Serial.begin(115200);
  delay(10);

  // prepare GPIO2
  pinMode(2, INPUT);

  // We start by connecting to a WiFi network
  Serial.println();
  Serial.println();
  Serial.print("Connecting to ");
  Serial.println(ssid);

  WiFi.begin(ssid, password);

  while (WiFi.status() != WL_CONNECTED) {
    delay(500);
    Serial.print(".");
  }

  Serial.println("");
  Serial.println("WiFi connected");
  Serial.println("IP address: ");
  Serial.println(WiFi.localIP());
}
```

- ▸ Delay for 5 seconds before starting the `loop` and then read the status of the GPIO2 pin:

```
void loop() {
  delay(5000);
  int val = digitalRead(2);
```

- ▸ Connect to the server:

```
  Serial.print("connecting to ");
  Serial.println(host);
```

```
// Use WiFiClient class to create TCP connections
WiFiClient client;
const int httpPort = 80;
if (!client.connect(host, httpPort)) {
  Serial.println("connection failed");
  return;
}
```

▶ Generate the URL for our GET request and print it on the serial monitor:

```
// We now create a URI for the request
String url = "/gpio/";
url += (val)?"1":"0";



Serial.print("Requesting URL: ");
Serial.println(url);
```

▶ Send the request to the server and check whether the connection has been timed out:

```
// This will send the request to the server
client.print(String("GET ") + url + " HTTP/1.1\r\n" +
"Host: " + host + "\r\n" +
"Connection: close\r\n\r\n");
  unsigned long timeout = millis();
  while (client.available() == 0) {
    if (millis() - timeout > 5000) {
      Serial.println(">>> Client Timeout !");
      client.stop();
      return;
    }
  }
```

▶ Read the reply from the server and display it on the serial monitor, then close the connection:

```
// Read all the lines of the reply from server and print them to
Serial
  while(client.available()){
    String line = client.readStringUntil('\r');
    Serial.print(line);
  }

  Serial.println();
  Serial.println("closing connection");
}
```

Upload this sketch to the ESP8266 board that will be used as a client. Remember to change `your_ssid` and `your_password` to the name for the access point SSID and password respectively, before uploading the code. Also replace `192.XXX.XXX.XXX` with the IP address of the ESP8266 board that is configured as a server.

Once you have uploaded the code onto both ESP8266 boards, turn on the setups. You can do that by connecting both ESP8266 boards to your computer via USB cable, or power them via external means (battery/power adapter). However, it is best to leave the ESP8266 client setup connected to the computer, so that you can read feedback from the server on the Arduino serial monitor. You will be able to turn on or off the LED connected to the ESP8266 server by pressing or releasing the **Push** button respectively.

 Turn on the ESP8266 server setup first, so that it can create the access point, before turning on the ESP8266 client.

How it works...

The ESP8266 server sketch starts by including the required ESP8266 libraries, after which the access point SSID and password are defined and the `Wi-Fi server` object created. In the setup section of the sketch, the program initializes the serial port and creates a Wi-Fi access point using the provided SSID and password. The server is initialized after that.

The ESP8266 server listens to see whether there are any clients that are connecting to it. If they are there, it reads the request that the client has sent and checks to see if it is valid. If the request contains the string `/gpio/0` and `/gpio/1`, it is considered valid.

A `/gpio/0` request causes the ESP8266 to turn off the LED and send back a reply to the client that says `LOW` to signify the state of GPIO2 pin. A `/gpio/1` request causes the ESP8266 to turn on the LED and send back a reply to the client that says `HIGH` to signify the state of the GPIO2 pin. Once the reply has been sent, the server ends the session and waits for another client to connect to it.

The ESP8266 client sketch also starts by including the `ESP8266 Wi-Fi` library and definition of the SSID, password, and host IP address of the ESP8266 server. The sketch then initializes the serial port and connects to the ESP8266 access point.

In the `loop` section of the sketch, the state of GPIO2 is read and the ESP8266 client connects to the server. If the state of the GPIO2 pin is high (**Push** button pressed), the client sends a `/gpio/1` request to the server. If the state of GPIO2 is low (**Push** button released), the client sends a `/gpio/0` request to the server. The ESP8266 then reads the reply from the server. If the data was sent to the server successfully, the ESP8266 client receives the state of the GPIO2 pin of the ESP8266 server.

See also

This recipe has shown you how to perform local M2M interactions using two ESP8266 boards. Having mastered that, you can now proceed to the next recipe and learn how to perform machine to machine interactions in the cloud.

Cloud machine to machine interaction

Interaction between two ESP8266 boards in remote locations is possible through cloud machine to machine communication. This recipe is going show you how to build a simple M2M project in the cloud with two ESP8266 boards. To achieve that, we will use MQTT.

MQTT is an M2M connectivity protocol that is used in the Internet of Things. It is a lightweight messaging transport that allows devices to publish or subscribe to different content online. Due to its lightweight nature, it uses very little bandwidth and resources while ensuring reliable data transmission between different clients.

The publish and subscribe pattern is what forms the basis for the MQTT protocol. The client that is sending a message is known as the publisher and the one that receives the message is called the subscriber. The publisher and subscriber do not interact directly with each other. There is a server known as the broker, which the publisher and subscriber connect to. The broker filters all the incoming messages and distributes them in the right order.

To communicate via MQTT, a device publishes data to an MQTT broker using a specific topic. For instance, if the device is publishing temperature data, it can publish it on a topic called temperature on MQTT. The device that is meant to receive the temperature data subscribes to that topic on the MQTT broker and receives temperature data as soon as it is published. We will demonstrate that using two ESP8266 boards.

Getting ready

You will need the following hardware components for this project:

- Two ESP8266 boards
- Two USB cables
- 220 Ω resistor
- LED
- Momentary **Push** button
- 10 kΩ resistor

1. The setup will resemble the one in the previous recipe. We will have one setup with **Push** button input and the other with an LED output.

2. To use MQTT on the ESP8266, you will need the `PubSubClient` library.

3. To install it, click on **Sketch|Include Library|Manage Libraries** on the Arduino IDE.

4. Then search for the `PubSubClient` library and install it:

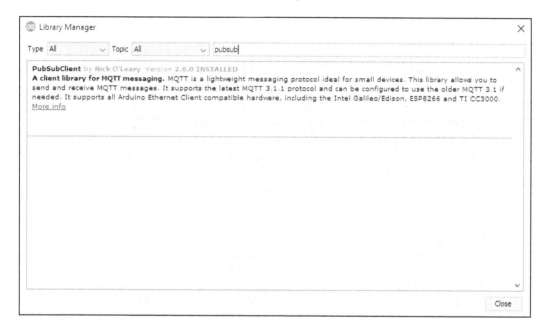

How to do it...

To successfully communicate via MQTT, we will set up one of the ESP8266 boards as the publisher and the other as the subscriber.There will be two topics that we are going to use for communication between the two ESP8266 boards. They are `buttonState` and `ledState`.

You will configure the ESP8266 board connected to the push button to publish data to the `buttonState` topic and subscribe to the `ledState` topic. Then configure the ESP8266 board connected to the LED, to publish data to the `ledState` topic and subscribe to the `buttonState` topic. This way, the ESP8266 boards will be able to communicate to each other directly, through the MQTT broker.

The sketch allows the publisher to send data to the MQTT broker when the **Push** button is pressed. The **Push** button will act as a toggle switch, such that pressing it changes the state of the LED connected to the subscriber ESP8266. The ESP8266 board that has an LED will be sending back the status of the LED to the ESP8266 with the **Push** button.

The sketch for the ESP8266 board with the **Push** button is as follows:

1. Include the libraries:

```
#include <ESP8266WiFi.h>
#include <PubSubClient.h>
```

2. Provide the SSID and `password` of your Wi-Fi network:

```
// Update these with values suitable for your network.
const char* ssid = "your_ssid";
const char* password = "your_password";
```

3. Provide the MQTT server and the topic that your device will be publishing to, and the topic that your device is subscribed to:

```
const char* mqtt_server = "broker.mqtt-dashboard.com";
const char* topicPub = "buttonState";
const char* topicSub = "ledState";
```

4. Create the `Wi-Fi client` and `MQTTclient` objects:

```
WiFiClient espClient;
PubSubClient client(espClient);
```

5. Variables to be used in the program:

```
long lastMsg = 0;// time last message was published
char msg[50]; // holds message that was published
int value = 0; // keeps track of number of publishes
boolean toggleState = false; // button toggles this state
boolean currentState = true; // current state of toggle
```

6. Configure the built-in LED pin as an output, initialize serial port, connect to the Wi-Fi hotspot, and then set up the MQTT server and `callback`:

```
void setup() {
  pinMode(BUILTIN_LED, OUTPUT);      // Initialize the BUILTIN_LED
pin as an output
  Serial.begin(115200);
  setup_wifi();
  client.setServer(mqtt_server, 1883);
  client.setCallback(callback);
}
```

7. Function that connects the ESP8266 board to the Wi-Fi hotspot:

```
void setup_wifi() {

  delay(10);
  // We start by connecting to a WiFi network
  Serial.println();
```

```
Serial.print("Connecting to ");
Serial.println(ssid);

WiFi.begin(ssid, password);

while (WiFi.status() != WL_CONNECTED) {
  delay(500);
  Serial.print(".");
}

Serial.println("");
Serial.println("WiFi connected");
Serial.println("IP address: ");
Serial.println(WiFi.localIP());
}
```

8. This is the `Callback` function that handles the incoming data from the topic the device has subscribed to:

```
void callback(char* topic, byte* payload, unsigned int length) {
  Serial.print("Message arrived [");
  Serial.print(topic);
  Serial.print("] ");
  for (int i = 0; i < length; i++) {
    Serial.print((char)payload[i]);
  }
  Serial.println();

}
```

9. Function that connects the ESP8266 client to the MQTT server and sets up the topic to `publish` to and the topic to subscribe to:

```
void reconnect() {
  // Loop until we're reconnected
  while (!client.connected()) {
    Serial.print("Attempting MQTT connection...");
    // Attempt to connect
    if (client.connect("ESP8266Client")) {
      Serial.println("connected");
      // Once connected, publish an announcement...
      client.publish(topicPub, "0");
      // ... and resubscribe
      client.subscribe(topicSub);
    } else {
      Serial.print("failed, rc=");
```

```
        Serial.print(client.state());
        Serial.println(" try again in 5 seconds");
        // Wait 5 seconds before retrying
        delay(5000);
      }
    }
  }
```

10. The `voidloop` section will run the previous functions. It will reconnect the ESP8266 to the MQTT when it disconnects and `publish` a message to the topic every 2 seconds:

```
void loop() {
// toggle the state if push button is pressed
if(digitalRead(2) && currentState == toggleState){
    toggleState = !toggleState;
  }
  if (!client.connected()) {
    reconnect(); //establish connection to MQTT
  }
  client.loop();

  long now = millis();
  if (now - lastMsg > 2000) { // 2 seconds have elapsed
    lastMsg = now;
    ++value;
snprintf (msg, 75, "%ld", toggleState); // compose message
    Serial.print("Publish message: ");
    Serial.println(msg);
    client.publish(topicPub, msg); // publish message
currentState = toggleState;
  }
}
```

The code for the ESP8266 board that has an LED connected to it will be as follows:

1. Include the libraries:

```
#include <ESP8266WiFi.h>
#include <PubSubClient.h>
```

2. Provide the SSID and `password` of your Wi-Fi network:

```
// Update these with values suitable for your network
const char* ssid = "your_ssid";
const char* password = "your_password";
```

3. Provide the MQTT server and the topic that your device will be publishing to, and the topic that your device is subscribed to:

```
const char* mqtt_server = "broker.mqtt-dashboard.com";
const char* topicPub = "ledState";
const char* topicSub = "buttonState";
```

4. Create the `Wi-Fi client` and `MQTTclient` objects:

```
WiFiClient espClient;
PubSubClient client(espClient);
```

5. Variables to be used in the program:

```
long lastMsg = 0;// time last message was published
char msg[50]; // holds message that was published
int value = 0; // keeps track of number of publishes
char incoming = '0'; // holds incoming character
```

6. Configure the built-in LED pin as an output, initialize serial port, connect to the Wi-Fi hotspot, and then set up the MQTT server and `callback`:

```
void setup() {
  pinMode(BUILTIN_LED, OUTPUT);      // Initialize the BUILTIN_LED
pin as an output
  Serial.begin(115200);
  setup_wifi();
  client.setServer(mqtt_server, 1883);
  client.setCallback(callback);
}
```

7. Function that connects the ESP8266 board to the Wi-Fi hotspot:

```
void setup_wifi() {

  delay(10);
  // We start by connecting to a WiFi network
  Serial.println();
  Serial.print("Connecting to ");
  Serial.println(ssid);

  WiFi.begin(ssid, password);

  while (WiFi.status() != WL_CONNECTED) {
    delay(500);
    Serial.print(".");
  }

  Serial.println("");
```

```
      Serial.println("WiFi connected");
      Serial.println("IP address: ");
      Serial.println(WiFi.localIP());
   }
```

8. `Callback` function that handles the incoming data from the topic the device has subscribed to:

```
void callback(char* topic, byte* payload, unsigned int length) {
   Serial.print("Message arrived [");
   Serial.print(topic);
   Serial.print("] ");
   for (int i = 0; i < length; i++) {
     Serial.print((char)payload[i]);
   }
   Serial.println();
incoming = payload[0];
   // Switch on the LED if an 1 was received as first character
   if ((char)payload[0] == '0') {
     digitalWrite(2, LOW);    // Turn the LED on (Note that LOW is
the voltage level
     // but actually the LED is on; this is because
     // it is acive low on the ESP-01)
   } else {
     digitalWrite(2, HIGH);   // Turn the LED off by making the
voltage HIGH
   }
}
```

9. Function that connects the ESP8266 client to the MQTT server and sets up the topic to `publish` to, and the topic to subscribe to:

```
void reconnect() {
  // Loop until we're reconnected
  while (!client.connected()) {
    Serial.print("Attempting MQTT connection...");
    // Attempt to connect
    if (client.connect("ESP8266Client")) {
      Serial.println("connected");
      // Once connected, publish an announcement...
      client.publish(topicPub, "low");
      // ... and resubscribe
      client.subscribe(topicSub);
    } else {
      Serial.print("failed, rc=");
```

```
        Serial.print(client.state());
        Serial.println(" try again in 5 seconds");
        // Wait 5 seconds before retrying
        delay(5000);
    }
  }
}
```

10. The `voidloop` section will run the previous functions. It will reconnect the ESP8266 to the MQTT when it disconnects, and `publish` a message to the topic every 2 seconds:

```
void loop() {
  if (!client.connected()) {
    reconnect(); //establish connection to MQTT
  }
  client.loop();

  long now = millis();
  if (now - lastMsg > 2000) { // 2 seconds have elapsed
    lastMsg = now;
    ++value;
  // publish low if incoming character is 0 and high if incoming
character is 1
    snprintf (msg, 75, incoming == '0'?"low":"high"); // compose
message
    Serial.print("Publish message: ");
    Serial.println(msg);
    client.publish(topicPub, msg); // publish message
  }
}
```

Replace your_ssid and your_password in the two sketches to match the credentials of your Wi-Fi network. Upload the sketches to their respective ESP8266 boards and power them. You can open the serial monitor to check the incoming data on each board.

How it works...

Both sketches start with including the ESP8266 `Wi-Fi library` and the `PubSubClient` library for MQTT. The Wi-Fi hotspot SSID and password, and the MQTT server, are defined and the publishing and subscription topics described. For the ESP8266 board that is connected to the push button, the publishing topic is `buttonState` and the subscription topic is `ledState`. For the ESP8266 board that is connected to the LED, the publishing and subscription topics are interchanged. A `Wi-Fi client` object and an `MQTT` object are created and the variables declared.

In the setup section of the sketches, the ESP8266 boards are connected to the Wi-Fi network whose credentials are provided. The serial port is initialized and the MQTT server started, then the callback action for incoming messages is initialized. After this stage, the sketches differ slightly for each setup.

For the ESP8266 **Push** button setup, the sketch reads the **Push** button input. If it is high (button is pressed), the sketch toggles the state that is going to be published to the MQTT server. If the previous state was high, it changes to low, and vice versa. Once the button input has been received, the sketch establishes a connection between the ESP8266 and MQTT server and sets the publishing topic and the subscription topic. Then if 2 seconds have elapsed since the last publish, the ESP8266 board publishes the state. The ESP8266 then waits for a response from the ESP8266 LED setup.

For the ESP8266 LED setup, the sketch is going to listen for any incoming data from the ESP8266 **Push** button setup. If the incoming character is a 0, the LED is turned off and if the incoming character is 1, the LED is turned on. The ESP8266 then sends a reply to the ESP8266 **Push** button setup indicating the state of the LED, that is, high if **ON** and low if **OFF**.

See also

You can also achieve M2M interaction between two ESP8266 boards using IFTTT. The next recipe will show you how to do that.

Automated M2M with IFTTT

In this recipe, we will look at how to use IFTTT for your M2M projects. To do so, we will use IFTTT to send data from one ESP8266 board to another. This is going to be facilitated by the IFTTT maker channel and an MQTT broker. We will use the Adafruit IO MQTT broker in this recipe.

Getting ready

You will need the following hardware components for this project:

- Two ESP8266 boards
- Two USB cables
- 220 Ω resistor
- LED
- Momentary **Push** button
- 10 kΩ resistor

The setup will resemble the one in *Basic local machine to machine interaction* of this chapter. We will have one ESP8266 board setup with a **Push** button input and the other with an LED output.

Install the `Adafruit MQTT` library. To do that, search for `adafruit mqtt` in the libraries manager and select the first result:

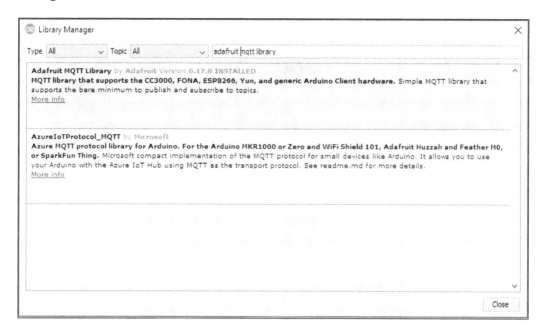

You will also need an IFTTT account at `https://ifttt.com` and an Adafruit IO account at `https://io.adafruit.com/`. You can refer back to *Chapter 6*, *Interacting with Web Services* for details on how to use IFTTT.

How to do it...

Refer the following steps:

1. Log in to Adafruit IO and click on **Settings** from the menu on the left side of the homepage. You are going to find all the credentials you need to add to your sketch to successfully establish a connection with the Adafruit MQTT broker. In this recipe, you will need the `AIO` key and your username. You will include these credentials in the code for your ESP8266 LED setup.

2. To generate the `AIO` key, click on the **View AIO Key** button and then click on **Generate AIO key**. This will generate a new `AIO` key for you. You can regenerate it later if you want to.

3. To get your username, click on the **Manage Account** button. This will list your account details, including your username. Record both credentials, because you will include them in your sketch.

4. The next step will be creating feeds. To do so, click on **Feeds** from the menu on the left side of the screen.

5. Click on **Actions** and create a feed called `stateToggle`. Once that is done, you can proceed to set up IFTTT:

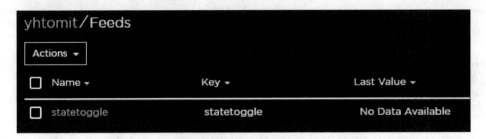

6. In IFTTT, select the maker channel as the trigger channel. Then set the event name to `toggle` and click on **Create trigger**:

7. Once that is done, select the Adafruit channel as the action channel. Provide the name of your feed in Adafruit IO and select the data to save. In this recipe, the name of our feed is `statetoggle` and the data to save will be `{{Value1}}`:

8. Click on **Create action** and open the page at `https://ifttt.com/maker`:

9. Click on **Settings** to get the key. You will need the key in the code for your ESP8266 **Push** button setup.

The code for the ESP8266 board with the LED setup is as follows:

1. Include the libraries:

```
#include <ESP8266WiFi.h>
#include "Adafruit_MQTT.h"
#include "Adafruit_MQTT_Client.h"
```

2. Define the SSID and password for your Wi-Fi hotspot:

```
#define WLAN_SSID        "your_ssid"
#define WLAN_PASS        "your_password"
```

3. Define the Adafruit IO credentials:

```
#define AIO_SERVER       "io.adafruit.com"
#define AIO_SERVERPORT   1883
#define AIO_USERNAME     "your_username"
#define AIO_KEY          "your_key"
```

4. Create a `Wi-Fi` client object, set up the `MQTT` server, and set up a feed:

```
// Create an ESP8266 WiFiClient class to connect to the MQTT
server.
WiFiClient client;

// Setup the MQTT client class by passing in the WiFi client and
MQTT server and login details.
Adafruit_MQTT_Client mqtt(&client, AIO_SERVER, AIO_SERVERPORT,
AIO_USERNAME, AIO_USERNAME, AIO_KEY);

// Setup a feed called 'statetoggle' for subscribing to changes to
the button
Adafruit_MQTT_Subscribe onoffbutton = Adafruit_MQTT_
Subscribe(&mqtt, AIO_USERNAME "/feeds/statetoggle", MQTT_QOS_1);
```

5. The `Callback` function gets incoming data and turns on and off the LED depending on the received data:

```
void onoffcallback(char *data, uint16_t len) {
  Serial.print("Hey we're in a onoff callback, the button value
is: ");
  Serial.println(data);
 if(*data == '0')digitalWrite(2, LOW);
   else digitalWrite(2, HIGH);
}
```

6. Initialize serial port, connect the ESP8266 to the Wi-Fi hotspot, set up the `callback` function, and subscribe to the feed on Adafruit IO:

```
void setup() {
  Serial.begin(115200);
  delay(10);
pinMode(2, OUTPUT);
  digitalWrite(2, LOW);
  Serial.println(F("Adafruit MQTT demo"));

  // Connect to WiFi access point.
  Serial.println(); Serial.println();
  Serial.print("Connecting to ");
  Serial.println(WLAN_SSID);

  WiFi.begin(WLAN_SSID, WLAN_PASS);
  while (WiFi.status() != WL_CONNECTED) {
    delay(500);
    Serial.print(".");
  }
```

```
  Serial.println();

  Serial.println("WiFi connected");
  Serial.println("IP address: "); Serial.println(WiFi.localIP());
// initialize callback
  onoffbutton.setCallback(onoffcallback);

// Setup MQTT subscription for feed.
  mqtt.subscribe(&onoffbutton);
}
```

7. Connect to the `MQTT` server and listen for any incoming packets:

```
void loop() {
  // Ensure the connection to the MQTT server is alive (this will
make the first
  // connection and automatically reconnect when disconnected).
See the MQTT_connect
  // function definition further below.
  MQTT_connect();

  // this is our 'wait for incoming subscription packets and
callback em' busy subloop
  // try to spend your time here:
  mqtt.processPackets(10000);

  // ping the server to keep the mqtt connection alive
  // NOT required if you are publishing once every KEEPALIVE
seconds

  if(! mqtt.ping()) {
    mqtt.disconnect();
  }
}
// Function to connect and reconnect as necessary to the MQTT
server.
// Should be called in the loop function and it will take care if
connecting.
void MQTT_connect() {
  int8_t ret;

  // Stop if already connected.
  if (mqtt.connected()) {
    return;
  }
```

```
        Serial.print("Connecting to MQTT... ");

        uint8_t retries = 3;
        while ((ret = mqtt.connect()) != 0) { // connect will return 0
    for connected
            Serial.println(mqtt.connectErrorString(ret));
            Serial.println("Retrying MQTT connection in 10
    seconds...");
            mqtt.disconnect();
            delay(10000);   // wait 10 seconds
            retries--;
            if (retries == 0) {
              // basically die and wait for WDT to reset me
              while (1);
            }
        }
        Serial.println("MQTT Connected!");
    }
```

The sketch for the ESP8266 that is connected to the **Push** button is as follows:

1. Include the library:

   ```
   #include <ESP8266WiFi.h>
   ```

2. Set the `time` between posts:

   ```
   #define timeInterval 10000 // time between each trigger
   ```

3. Set the Wi-Fi hotspot credentials:

   ```
   const char* ssid     = "your_ssid";
   const char* password = "your_password";
   ```

4. Define the IFTTT credentials:

   ```
   const char* host = "maker.ifttt.com";
   const char* privateKey = "your_key";
   const char* event = "toggle";
   ```

5. Variables:

   ```
   boolean toggleState = false; // holds state of button
   boolean currentState = false; // holds current button state
   long lastTime = 0; // holds previous time trigger was sent
   ```

6. Initialize serial port and connect to the Wi-Fi hotspot:

   ```
   void setup() {
     Serial.begin(115200); // initialize serial communication
     pinMode(2, INPUT);
   ```

```
    delay(100);

    // We start by connecting to a WiFi network

    Serial.println();
    Serial.println();
    Serial.print("Connecting to ");
    Serial.println(ssid);

    WiFi.begin(ssid, password);

    while (WiFi.status() != WL_CONNECTED) {
      delay(500);
      Serial.print(".");
    }

    Serial.println("");
    Serial.println("WiFi connected");
    Serial.println("IP address: ");
    Serial.println(WiFi.localIP());
}
```

7. Read the **Push** button state:

```
void loop() {
  if(digitalRead(2) && currentState == toggleState){
    toggleState = !toggleState;
    Serial.println("pressed");
  }
```

8. If the time interval has passed and the button has been pressed, connect to IFTTT:

```
  if(currentState != toggleState && millis()- lastTime >
timeInterval){ // temperature is less than 30 deg celsius
    Serial.print("connecting to ");
    Serial.println(host);
    // Use WiFiClient class to create TCP connections
    WiFiClient client;
    const int httpPort = 80;
    if (!client.connect(host, httpPort)) {
      Serial.println("connection failed");
      return;
    }
```

9. Generate the URL:

```
// We now create a URI for the request
String url = "/trigger/";
url += event;
url += "/with/key/";
url += privateKey;
url += "?value1=";
url += String(toggleState);

Serial.print("Requesting URL: ");
Serial.println(url);
```

10. Send a GET request and the read incoming data from the server:

```
// This will send the request to the server
client.print(String("GET ") + url + " HTTP/1.1\r\n" +
"Host: " + host + "\r\n" +
"Connection: close\r\n\r\n");
    unsigned long timeout = millis();
    while (client.available() == 0) {
      if (millis() - timeout > 5000) {
        Serial.println(">>> Client Timeout !");
        client.stop();
        return;
      }
    }

    // Read all the lines of the reply from server and print them
to Serial
    while(client.available()){
      String line = client.readStringUntil('\r');
      Serial.print(line);
    }
    Serial.println();
    Serial.println("closing connection");
    lastTime = millis(); // save time of last trigger
    currentState = toggleState;
  }
}
```

Before uploading the code, replace `your_ssid` in the sketch with the SSID of your Wi-Fi network and `your_password` with the password of your Wi-Fi network. Upload the code and open the serial monitor to check the status messages.

How it works...

The ESP8266 **Push** button setup acts as the input to the system. When you press the **Push** button, it toggles the state of a `Boolean` variable called `toggleState`. The value of the `Boolean` variable is then passed to IFTTT through the maker channel toggle event that you created. This is known as triggering. When triggered, the toggle event on the maker channel passes the value of the `Boolean` variable to the Adafruit IO channel, which posts the value to the feed that you specified in the settings. The feed in this recipe was called `statetoggle`.

The ESP8266 LED setup is the output of the system. It subscribes to the `statetoggle` feed in the Adafruit IO MQTT broker and listens for any incoming data. If there is any data that has been posted to the feed via IFTTT, the ESP8266 board reads it and changes the LED output state to match the incoming character. When the incoming character is 1, the LED is turned on, and when it is 0, the LED is turned off.

See also

Now that you have learned the different ways of achieving M2M interaction between ESP8266 boards, you should be able to implement a few projects using M2M interaction. The next recipe is going to show you how to create a simple alarm system using M2M communication.

M2M alarm system

In the previous chapters, we looked at the different ways of achieving M2M communication between two ESP8266 boards. In this recipe, we will look at how to implement M2M communication in an IoT project. To do that, we will create a simple M2M alarm system with one ESP8266 board connected to a DHT11 sensor and the other connected to a buzzer. The alarm will be triggered when the temperature reading drops below 20 degrees Celsius.

Getting ready

You will need the following hardware components:

- ESP8266 board
- USB cable
- DHT11 temperature/humidity sensor (https://www.sparkfun.com/products/10167)
- 10 kΩ resistor
- Buzzer
- Breadboard
- Jumper wires

The DHT11 pin configuration is shown in the following figure:

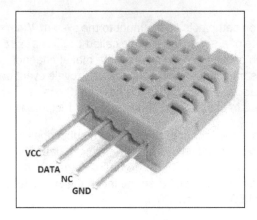

1. First, mount the ESP8266 board and the DHT11 sensor onto the breadboard.

2. Connect a 10 kΩ pull up resistor to the DHT11 data pin and connect the VCC pin and GND pin to the 3V pin and GND pin of the ESP8266 board, respectively.

3. Finally, connect the data pin of the DHT11 to GPIO2 of the ESP8266 board. Use jumper wires to do the connections.

The setup is shown in the following figure:

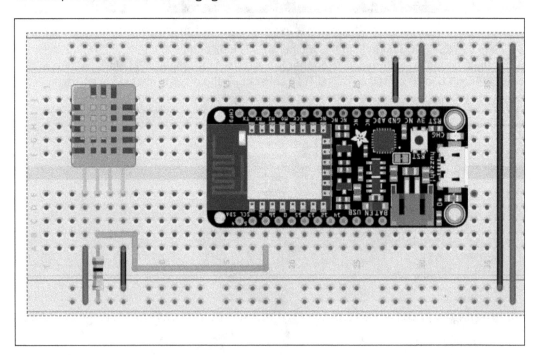

For the second setup, we will have the ESP8266 GPIO2 connected to a buzzer. It will act as the siren for the alarm system. The setup is as shown here:

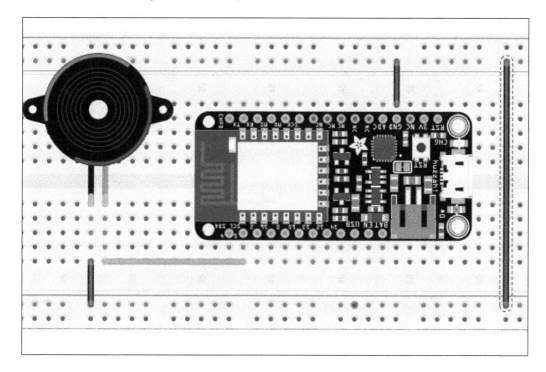

How to do it...

Configure the ESP8266 board connected to the DHT11 sensor (trigger) as a Wi-Fi client, and the ESP8266 board connected to the buzzer (alarm) as a Wi-Fi server and access point. This way, the trigger setup is going to connect to the access point created by the alarm setup, and send a HTTP request to the alarm setup every time the temperature goes down and above 20 degree Celsius.

There will be two kinds of trigger that will be sent to the alarm setup. The first trigger will be /gpio/0, which will turn off the alarm. This trigger will be sent to the alarm setup when the temperature goes above 20 degrees Celsius. The second trigger will be /gpio/1, which will turn on the alarm. This trigger will be sent to the alarm setup when the temperature goes below 20 degrees Celsius.

The code for the alarm setup is as follows:

1. Include libraries:

```
#include <ESP8266WiFi.h>
#include <WiFiClient.h>
#include <ESP8266WebServer.h>
```

2. Set the SSID and `password` for your access point:

```
const char* ssid = "hotspot_ssid";
const char* password = "hotspot_password";

// Create an instance of the server
// specify the port to listen on as an argument
WiFiServer server(80);
```

3. Initialize the serial communication port:

```
void setup() {
  delay(1000);
  Serial.begin(115200);
  delay(10);
```

4. Set `GPIO2` as output:

```
  // prepare GPIO2 where buzzer is connected
  pinMode(2, OUTPUT);
  digitalWrite(2, 0);
```

5. Start the access point using the SSID and `password` that you provided earlier:

```
  Serial.print("Configuring access point...");
  /* You can remove the password parameter if you want the AP to
be open. */
  WiFi.softAP(ssid, password);

  IPAddress myIP = WiFi.softAPIP();
  Serial.print("AP IP address: ");
  Serial.println(myIP);

  server.begin();
  Serial.println("HTTP server started");
}
```

6. Check the `client` has connected to the server and read any incoming data from the client:

```
void loop() {
  // Check if a client has connected
  WiFiClient client = server.available();
  if (!client) {
    return;
  }

  // Wait until the client sends some data
  Serial.println("new client");
```

```
  while(!client.available()){
    Serial.print('.');
    delay(1);
  }

  // Read the first line of the request
  String req = client.readStringUntil('\r');
  Serial.println(req);
  client.flush();
```

7. Evaluate the `request` from the `client` to determine what state the GPIO2 pin should be set at:

```
  // Match the request
  int val;
  if (req.indexOf("/gpio/0") != -1)
    val = 0;
  else if (req.indexOf("/gpio/1") != -1)
    val = 1;
  else {
    Serial.println("invalid request");
    client.stop();
    return;
  }
```

8. Set the GPIO2 pin state to correspond to the `request` from the client. This will turn the buzzer on or off, depending on the value of the `val` variable:

```
  // Set GPIO2 according to the request
// turns on or off buzzer
  digitalWrite(2, val);

  client.flush();
```

9. Generate the `response` that the server will send to the client:

```
  // Prepare the response
  String s = (val)?"high":"low";
```

10. Send the response to the `client` and end the session:

```
  // Send the response to the client
  client.print(s);
  delay(1);
  Serial.println("Client disonnected");

  // The client will actually be disconnected
  // when the function returns and 'client' object is detroyed
}
```

11. Upload this sketch to the ESP8266 board that will be used as a server. Remember to change `hotspot_ssid` and `hotspot_password` to your preferred access point SSID and password respectively, before uploading the code.

12. Open the serial monitor and note down the IP address of your ESP8266 server. You will include it in the ESP8266 trigger sketch.

The trigger setup sketch is as follows:

1. Include library:

   ```
   #include <ESP8266WiFi.h>
   #include "DHT.h"
   ```

2. Set up the DHT11 sensor:

   ```
   #define DHTPIN 2      // what digital pin we're connected to
   #define DHTTYPE DHT11    // sensor type - DHT 11
   DHT dht(DHTPIN, DHTTYPE); // create DHT object
   ```

3. Set the SSID and `password` for your server's access point:

   ```
   const char* ssid     = "your-ssid";
   const char* password = "your-password";
   ```

4. Set the IP address of your ESP8266 server:

   ```
   const char* host = "192.XXX.XXX.XXX";
   ```

5. Declare some variables that will be used:

   ```
   boolean sent = false; // signifies request has been made
   int state = 0; // holds previous trigger state
   ```

6. Initialize the serial communication port and DHT11 sensor, set GPIO2 as an input, and connect to the Wi-Fi access point:

   ```
   void setup() {
     Serial.begin(115200);
    dht.begin();
     delay(10);

     // prepare GPIO2
     pinMode(2, INPUT);

     // We start by connecting to a WiFi network
     Serial.println();
     Serial.println();
     Serial.print("Connecting to ");
     Serial.println(ssid);
   ```

```
    WiFi.begin(ssid, password);

    while (WiFi.status() != WL_CONNECTED) {
      delay(500);
      Serial.print(".");
    }

    Serial.println("");
    Serial.println("WiFi connected");
    Serial.println("IP address: ");
    Serial.println(WiFi.localIP());
}
```

7. Delay for 5 seconds before starting the `loop` and then read the temperature from the DHT11 sensor. If the temperature is below `20` degrees Celsius, set the trigger as `1`, and if it is greater than or equal to `20` degrees Celsius, set the trigger as `0`:

```
void loop() {
  delay(5000);
// Read temperature as Celsius (the default)
  float t = dht.readTemperature();

  // Check if any reads failed and exit early (to try again).
  if (isnan(t)) {
    Serial.println("Failed to read from DHT sensor!");
    return;
  }
 // temp below 20 degC and previous trigger is low (0)
  if(t < 20 && state == 0){
    state = 1; // set current trigger high
    sent = false; // allow client to send http request
  }
  else if(t >= 20 && state == 1){ // temp < 20degC trigger = 1
    state = 0; // set current trigger low
    sent = false; // allow client to send http request
  }
```

8. Connect to the server if the `Boolean` variable sent is `false`:

```
if(!sent){
    Serial.print("connecting to ");
    Serial.println(host);

    // Use WiFiClient class to create TCP connections
    WiFiClient client;
    const int httpPort = 80;
```

```
      if (!client.connect(host, httpPort)) {
        Serial.println("connection failed");
        return;
      }
```

9. Generate the URL for our GET request and print it on the serial monitor:

```
    // We now create a URI for the request
      String url = "/gpio/";
      url += (state)?"1":"0";



      Serial.print("Requesting URL: ");
      Serial.println(url);
```

10. Send request to the server and check whether the connection has been timed out:

```
    // This will send the request to the server
      client.print(String("GET ") + url + " HTTP/1.1\r\n" +
    "Host: " + host + "\r\n" +
    "Connection: close\r\n\r\n");
      unsigned long timeout = millis();
      while (client.available() == 0) {
        if (millis() - timeout > 5000) {
          Serial.println(">>> Client Timeout !");
          client.stop();
          return;
        }
      }
```

11. Read the reply from the server and display it on the serial monitor, then close the connection:

```
    // Read all the lines of the reply from server and print them to
    Serial
      while(client.available()){
        String line = client.readStringUntil('\r');
        Serial.print(line);
      }

      Serial.println();
      Serial.println("closing connection");
    sent = true; // prevent request from being sent
      }
    }
```

12. Upload this sketch to the ESP8266 board that will be used as a server.

13. Remember to change `your_ssid` and `your_password` to the name for the access point SSID and `password` respectively, before uploading the code.

 Turn on the ESP8266 server setup first, so that it can set up the access point, before you turn on the ESP8266 client.

How it works...

The ESP8266 alarm setup will be the server and also the Wi-Fi access point. The ESP8266 trigger setup will be the client and will connect to the access point created by the ESP8266 alarm setup.

When the temperature reading from the DHT11 goes below 20 degrees Celsius, the ESP8266 will send an HTTP request containing the alarm trigger to the alarm setup. When the trigger is sent, a `Boolean` variable called `sent` is set to `true`, to prevent the trigger from being sent again, until the temperature rises to 20 degrees Celsius or above.

If the temperature rises to 20 degrees Celsius or above, the ESP8266 sends an HTTP request containing an off trigger to the alarm setup. The `Boolean` variable sent is set to true again to prevent the HTTP request from being sent again until the temperature goes below 20 degrees Celsius. When the alarm setup receives the HTTP request, it checks to see what kind of trigger it contains. If it is `/gpio/0`, the buzzer is turned off. If it is `/gpio/1`, the buzzer is turned on.

See also

Automated light control can be achieved using M2M. Proceed to the next chapter to see how to do it.

Automated light control

In this recipe, we will look at automated light control using M2M. We will work on a project that will turn a light source on or off, depending on the light intensity of the surrounding environment. Two ESP8266 boards will be used to accomplish this. One of them will have a light dependent resistor, and the other will have a relay connected to it.

You will need the following hardware components:

- ESP8266 board
- USB cable
- **Light dependent resistor (LDR)** (https://www.sparkfun.com/products/9088)
- 10 kΩ resistor
- Relay module (http://www.dx.com/p/arduino-5v-relay-module-blue-black-121354)
- Breadboard
- Jumper wires

Connect the analog input of the ESP8266 board to an LDR connected to a 10 kΩ resistor in a voltage divider circuit, as shown in the following image. This setup is going to act as the input for the system:

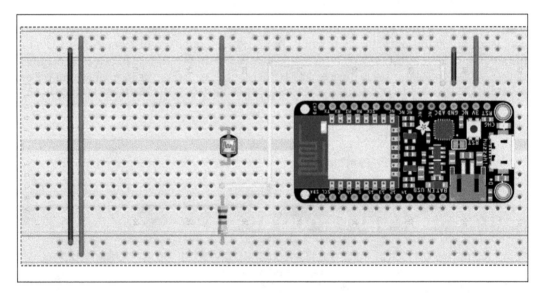

As for the output of the system, connect a relay to GPIO2 of the other ESP8266 board. The connection will be via an NPN transistor such as the 2N222, and a diode such as 1N4001 will be placed across the relay coil pins to protect the board from back EMF. The relay circuit should look like this:

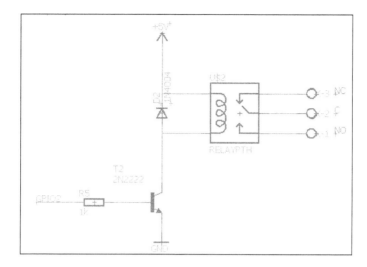

However, in this recipe, you won't have to connect an NPN transistor and the diode in your setup, as they are already provided on the relay module that you are going to use. You will connect three wires from your ESP8266 board to the relay module. The wires will be GPIO2, GND, and USB power (5V). This will simplify the hardware connection, as shown in the following figure:

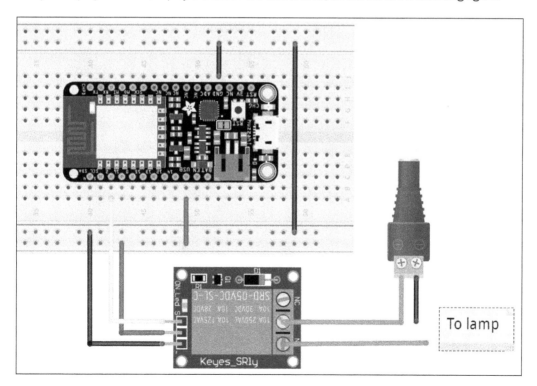

How to do it...

Configure the input setup, which comprises the LDR sensor and an ESP8266 board, to monitor the light intensity in the surroundings. If the reading on the analog pin goes below 512, the ESP8266 should send the value 1 to the MQTT broker under a topic called lightControl. If the value goes above 512, the ESP8266 board should send the value 0 to the MQTT broker under the same topic.

In addition to that, set up the ESP8266 to subscribe to a topic called `relayState` that will return the state of the relay switch from the output setup.

The code for the sensor setup is as follows:

1. Include the libraries:

```
#include <ESP8266WiFi.h>
#include <PubSubClient.h>
```

2. Provide the SSID and `password` of your Wi-Fi network:

```
// Update these with values suitable for your network.
const char* ssid = "your_ssid";
const char* password = "your_password";
```

3. Provide the MQTT server and the topic that your device will be publishing to, and the topic that your device is subscribed to:

```
const char* mqtt_server = "broker.mqtt-dashboard.com";
const char* topicPub = "lightControl";
const char* topicSub = "relayState";
```

4. Create the `Wi-Fi client` and `MQTT client` objects:

```
WiFiClient espClient;
PubSubClient client(espClient);
```

5. Variables to be used in the program:

```
char msg[50];
boolean triggerState = false; // type of trigger 0 or 1
boolean sent = false; // controls when to publish
```

6. Initialize serial port, connect to the Wi-Fi hotspot and then set up the MQTT server and `callback`:

```
void setup() {
  Serial.begin(115200);
  setup_wifi();
  client.setServer(mqtt_server, 1883);
  client.setCallback(callback);
}
```

7. Function that connects the ESP8266 board to the Wi-Fi hotspot:

```
void setup_wifi() {

  delay(10);
  // We start by connecting to a WiFi network
  Serial.println();
  Serial.print("Connecting to ");
  Serial.println(ssid);

  WiFi.begin(ssid, password);

  while (WiFi.status() != WL_CONNECTED) {
    delay(500);
    Serial.print(".");
  }

  Serial.println("");
  Serial.println("WiFi connected");
  Serial.println("IP address: ");
  Serial.println(WiFi.localIP());
}
```

8. This is the `Callback` function that handles the incoming data from the topic the device has subscribed to:

```
void callback(char* topic, byte* payload, unsigned int length) {
  Serial.print("Message arrived [");
  Serial.print(topic);
  Serial.print("] ");
  for (int i = 0; i < length; i++) {
    Serial.print((char)payload[i]);
  }
  Serial.println();
}
```

9. Function that connects the ESP8266 client to the MQTT server and sets up the topic to `publish` to, and the topic to subscribe to:

```
void reconnect() {
  // Loop until we're reconnected
  while (!client.connected()) {
    Serial.print("Attempting MQTT connection...");
    // Attempt to connect
    if (client.connect("ESP8266Client")) {
      Serial.println("connected");
      // Once connected, publish an announcement...
```

```
            client.publish(topicPub, "0");
            // ... and resubscribe
            client.subscribe(topicSub);
        } else {
            Serial.print("failed, rc=");
            Serial.print(client.state());
            Serial.println(" try again in 5 seconds");
            // Wait 5 seconds before retrying
            delay(5000);
        }
    }
}
```

10. Check sensor input and set `trigger` to the correct state:

```
void loop() {
  // toggle the state if push button is pressed
  if(analogRead(A0) < 512 && !triggerState){
      triggerState = true; // set trigger high
      sent = false; // allow publishing
  }
    else if(analogRead(A0) >= 512 && triggerState){
      triggerState = false; // set trigger low
      sent = false; // allow publishing
  }
```

11. Connect to the MQTT broker if `connection` has been lost:

```
    if (!client.connected()) {
        reconnect(); //establish connection to MQTT
    }
    client.loop();
```

12. Publish `trigger` if its state has changed:

```
if (!sent) {
    snprintf (msg, 50, "%ld", triggerState); // generate message
    Serial.print("Publish message: ");
    Serial.println(msg);
    client.publish(topicPub, msg); // publish
    sent = true;
  }
}
```

Configure the output setup to subscribe to the `lightControl` topic on the MQTT server. It will be listening for any incoming data from the input setup. The incoming data will either be a 1 or a 0. If it is a 1, the ESP8266 board in the output setup will set the GPIO2 pin high to close the relay. If the incoming data is 0, the ESP8266 will set the GPIO2 pin low, opening the relay. Configure the output setup to publish the current state of the relay to a topic called `relayState`. The input setup will be subscribed to this topic for updates on the current state of the relay. The code will look like this:

1. Include the libraries:

```
#include <ESP8266WiFi.h>
#include <PubSubClient.h>
```

2. Provide the SSID and `password` of your Wi-Fi network:

```
// Update these with values suitable for your network.
const char* ssid = "your_ssid";
const char* password = "your_password";
```

3. Provide the MQTT server and the topic that your device will be publishing to, and the topic that your device is subscribed to:

```
const char* mqtt_server = "broker.mqtt-dashboard.com";
const char* topicPub = " relayState";
const char* topicSub = " lightControl";
```

4. Create the `Wi-Fi client` and `MQTT client` objects:

```
WiFiClient espClient;
PubSubClient client(espClient);
```

5. Variables to be used in the program:

```
char msg[50];
char incoming = '0'; // holds incoming character
boolean sent = false; // controls publishing
```

6. Configure the built-in LED pin as an output, initialize serial port, connect to the Wi-Fi hotspot, and then set up the MQTT server and `callback`:

```
void setup() {
pinMode(2, OUTPUT);      // Initialize GPIO2 pin as an output
   digitalWrite(2, LOW); // set pin low
   Serial.begin(115200);
   setup_wifi();
   client.setServer(mqtt_server, 1883);
   client.setCallback(callback);
}
```

7. Function that connects the ESP8266 board to the Wi-Fi hotspot:

```
void setup_wifi() {

  delay(10);
  // We start by connecting to a WiFi network
  Serial.println();
  Serial.print("Connecting to ");
  Serial.println(ssid);

  WiFi.begin(ssid, password);

  while (WiFi.status() != WL_CONNECTED) {
    delay(500);
    Serial.print(".");
  }

  Serial.println("");
  Serial.println("WiFi connected");
  Serial.println("IP address: ");
  Serial.println(WiFi.localIP());
}
```

8. `Callback` function that handles the incoming data from the topic the device has subscribed to:

```
void callback(char* topic, byte* payload, unsigned int length) {
  Serial.print("Message arrived [");
  Serial.print(topic);
  Serial.print("] ");
  for (int i = 0; i < length; i++) {
    Serial.print((char)payload[i]);
  }
  Serial.println();
  // Switch on the LED if an 1 was received as first character
  if ((char)payload[0] == '0'&& incoming != (char)payload[0]) {
    digitalWrite(2, LOW);    // open relay
    incoming = (char)payload[0];
    sent = false; // allow publishing
  } else if ((char)payload[0] == '1'&& incoming != (char)
payload[0]) {
    digitalWrite(2, HIGH);   // close relay
    incoming = (char)payload[0];
    sent = false; // allow publishing
  }
}
```

9. Function that connects the ESP8266 `client` to the MQTT server and sets up the topic to `publish` to, and the topic to subscribe to:

```
void reconnect() {
  // Loop until we're reconnected
  while (!client.connected()) {
    Serial.print("Attempting MQTT connection...");
    // Attempt to connect
    if (client.connect("ESP8266Client")) {
      Serial.println("connected");
      // Once connected, publish an announcement...
      client.publish(topicPub, "low");
      // ... and resubscribe
      client.subscribe(topicSub);
    } else {
      Serial.print("failed, rc=");
      Serial.print(client.state());
      Serial.println(" try again in 5 seconds");
      // Wait 5 seconds before retrying
      delay(5000);
    }
  }
}
```

10. The `voidloop` section will run the previous functions. It will reconnect the ESP8266 to the MQTT when it disconnects and `publish` a message to the topic every time the incoming data changes:

```
void loop() {
 if (!client.connected()) {
    reconnect(); //establish connection to MQTT
  }
  client.loop();

if (!sent) {
    snprintf (msg, 50, incoming == '0'?"OFF":"ON");
    Serial.print("Publish message: ");
    Serial.println(msg);
    client.publish(topicPub, msg);
    sent = true;
  }
}
```

Replace `your_ssid` and `your_password` in the code to match the credentials of your Wi-Fi network. Upload the sketches to their respective ESP8266 boards and power them.

How it works...

The ESP8266 input setup monitors light intensity using an LDR sensor. When the light in the surrounding environment is dim, indicating that night is approaching, the input setup sends a trigger to a topic called `lightControl` in an MQTT broker. The output setup reads data from that topic and turns the light on or off, depending on the received trigger. When the trigger is 0, the bulb is turned off, and when it is 1, the bulb is turned on.

There's more...

Change the sketch so that the lights go on when it is bright outside and go off when it is dim outside.

See also

Now that you have successfully created some M2M projects, it is time to challenge yourself with a more elaborate project. The next recipe will take you through creating an *Automated gardening controller*.

Automated gardening controller

M2M forms the basis for elaborate control systems that involve monitoring remote sensors. To demonstrate that, we will build an automated gardening controller using M2M. This will involve monitoring the soil moisture and temperature of a garden, and controlling a water pump and a ventilation system.

Getting ready

You will need the following hardware components for this project:

- Two ESP8266 boards
- Two USB cables
- Soil moisture sensor (`https://www.sparkfun.com/products/13322`)
- 220 Ω resistor
- 10 Ω resistor
- DHT11 (`https://www.sparkfun.com/products/10167`)

- 10 kΩ resistor
- Relay module (http://www.dx.com/p/arduino-5v-relay-module-blue-black-121354)
- Water pump (https://www.sparkfun.com/products/10398)
- 12V power adapter/source
- TIP120 NPN transistor
- 1 kΩ resistor

The first setup will be the sensor hub, which will comprise of an ESP8266 board, a soil moisture sensor, and a DHT11 sensor. Connect a 10 kΩ pull up resistor to the DHT11 data pin, and connect the VCC pin and GND pin of the sensor to the 3V pin and GND pin of the ESP8266 board, respectively. Finally connect the data pin of the DHT11 to GPIO #2 of the ESP8266 board. Now you can set up the soil moisture sensor.

Begin by connecting the soil moisture sensor VCC and GND pins to the ESP8266 board 3V and GND pins respectively. Then connect the SIG pin to the voltage divider. The voltage divider will be constructed using the 220 Ω and 100 Ω resistors. Connect the output of the voltage divider to the analog pin. The voltage divider schematic diagram is shown in the following image. The voltage divider setup reduces the soil moisture sensor output from 0V-3V to approximately 0V-1V, which is ideal for the ESP8266 analog to digital converter:

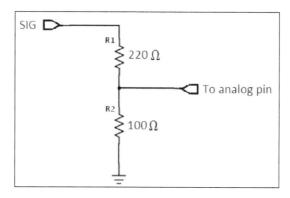

The complete setup will look like this:

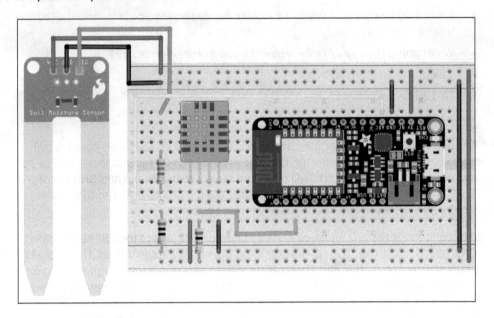

The actuator hub will consist of an ESP8266 board, a relay module, and a 12V vacuum water pump. Start by connecting the relay module signal pin to GPIO4 of the ESP8266 board, and the − and + pins of the relay module to the GND and 3V pins of the ESP8266 respectively. Connect the 12V vacuum pump positive pin to the +12V power supply, and the negative pin to the collector of a TIP120 NPN transistor. Connect the emitter of the NPN transistor to the GND pin of the ESP8266 board, and the base to the GPIO5 pin via a 1 kΩ resistor.

The setup will look like this:

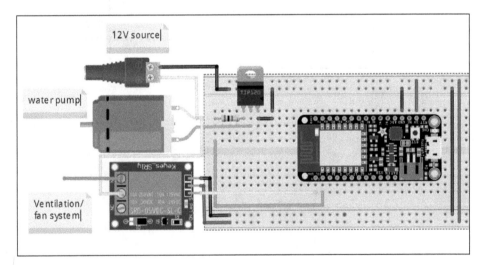

You will also need an IFTTT account on `https://ifttt.com` and an Adafruit IO account on `https://io.adafruit.com/`. You can refer back to *Chapter 6, Interacting with Web Services* for more details on how to use IFTTT.

How to do it...

Refer to the following steps:

1. Create two new feeds on Adafruit IO.

2. Name the feeds `moistureLog` and `temperatureLog`. Remember to copy the AIO key and username, as you will include them in the actuator hub code. Once that is done, you can set up IFTTT.

3. In IFTTT, select the maker channel as the trigger channel. Then set the event name as `moistureLog` and click on **Create trigger**:

4. Once that is done, select the Adafruit channel as the action channel. Select the name of your feed in Adafruit IO and select the data to save. For this applet, you will select the `moistureLog` feed, and the data to save will be `{{Value1}}`:

5. Click on **Create action** and open this page: `https://ifttt.com/maker`. Click on **Settings** to get the key. You will include this key in the sensor hub code.

6. Create one more IFTTT applet in the same way and give the maker channel trigger this event name: `temperatureLog`. The Adafruit IO feed that will correspond to that trigger event is `temperatureLog`.

The code for the ESP8266 board with the actuator hub is as follows:

1. Include the libraries:

```
#include <ESP8266WiFi.h>
#include "Adafruit_MQTT.h"
#include "Adafruit_MQTT_Client.h"
```

2. Define the SSID and `password` for your Wi-Fi hotspot:

```
#define WLAN_SSID        "your_ssid"
#define WLAN_PASS        "your_password"
```

3. Define the Adafruit IO credentials:

```
#define AIO_SERVER        "io.adafruit.com"
#define AIO_SERVERPORT    1883
#define AIO_USERNAME      "your_username"
#define AIO_KEY           "your_key"
```

4. Create a `Wi-Fi client` object, set up the MQTT server, and set up the feeds:

```
// Create an ESP8266 WiFiClient class to connect to the MQTT
server.
WiFiClient client;

// Setup the MQTT client class by passing in the WiFi client and
MQTT server and login details.
Adafruit_MQTT_Client mqtt(&client, AIO_SERVER, AIO_SERVERPORT,
AIO_USERNAME, AIO_USERNAME, AIO_KEY);


// Setup a feed called temperatureLog
Adafruit_MQTT_Subscribe temperature = Adafruit_MQTT_
Subscribe(&mqtt, AIO_USERNAME "/feeds/temperatureLog", MQTT_
QOS_1);

// Setup a feed called moistureLog
Adafruit_MQTT_Subscribe moisture = Adafruit_MQTT_Subscribe(&mqtt,
AIO_USERNAME "/feeds/moistureLog", MQTT_QOS_1);
```

5. These `Callback` functions get incoming data and turn on and off the corresponding outputs, depending on the received data:

```
//callback function for temperatureLog feed
void tempSwitchCallback(char *data, uint16_t len) {
  Serial.print("The relay trigger is: ");
  Serial.println(data);
  if(*data == '0')digitalWrite(4, LOW);
  else digitalWrite(4, HIGH);
}
//callback function for moistureLog feed
void moistureSwitchCallback(char *data, uint16_t len) {
  Serial.print("The motor trigger is: ");
  Serial.println(data);
  if(*data == '0')digitalWrite(5, LOW);
  else digitalWrite(5, HIGH);
}
```

6. Initialize serial port, configure the output GPIO pins and connect the ESP8266 to the Wi-Fi hotspot, set up the `callback` functions, and subscribe to the feeds on Adafruit IO:

```
void setup() {
  Serial.begin(115200);
  delay(10);
// set pin 4 and 5 as outputs
pinMode(4, OUTPUT);
pinMode(5, OUTPUT);
// initialize the pins in the low state
  digitalWrite(4, LOW);
digitalWrite(5, LOW);
  Serial.println(F("Adafruit MQTT demo"));

  // Connect to WiFi access point.
  Serial.println(); Serial.println();
  Serial.print("Connecting to ");
  Serial.println(WLAN_SSID);

  WiFi.begin(WLAN_SSID, WLAN_PASS);
  while (WiFi.status() != WL_CONNECTED) {
    delay(500);
    Serial.print(".");
  }
  Serial.println();

  Serial.println("WiFi connected");
Serial.println("IP address: ");
Serial.println(WiFi.localIP());

// initialize callbacks
temperature.setCallback(tempSwitchCallback);
  moisture.setCallback(moistureSwitchCallback);

  // Setup MQTT subscription for feeds
  mqtt.subscribe(&temperature);
  mqtt.subscribe(&moisture);}
```

7. Connect to the MQTT server and listen for any incoming packets:

```
void loop() {
  // Ensure the connection to the MQTT server is alive (this will
make the first
  // connection and automatically reconnect when disconnected).
See the MQTT_connect
  // function definition further below.
  MQTT_connect();
```

```
  // this is our 'wait for incoming subscription packets and
callback em' busy subloop
  // try to spend your time here:
  mqtt.processPackets(10000);

  // ping the server to keep the mqtt connection alive
  // NOT required if you are publishing once every KEEPALIVE
seconds

  if(! mqtt.ping()) {
    mqtt.disconnect();
  }
}
// Function to connect and reconnect as necessary to the MQTT
server.
// Should be called in the loop function and it will take care if
connecting.
void MQTT_connect() {
  int8_t ret;

  // Stop if already connected.
  if (mqtt.connected()) {
    return;
  }

  Serial.print("Connecting to MQTT... ");

  uint8_t retries = 3;
  while ((ret = mqtt.connect()) != 0) { // connect will return 0
for connected
      Serial.println(mqtt.connectErrorString(ret));
      Serial.println("Retrying MQTT connection in 10
seconds...");
      mqtt.disconnect();
      delay(10000);  // wait 10 seconds
      retries--;
      if (retries == 0) {
        // basically die and wait for WDT to reset me
        while (1);
      }
  }
  Serial.println("MQTT Connected!");
}
```

The sketch for the ESP8266 sensor hub is as follows:

1. Include the library:

```
#include <ESP8266WiFi.h>
#include "DHT.h"
```

2. Configure the `DHT11` sensor parameters:

```
#define DHTPIN 2       // what digital pin we're connected to
#define DHTTYPE DHT11   // sensor type - DHT 11
DHT dht(DHTPIN, DHTTYPE); // create DHT object
```

3. Set the Wi-Fi hotspot credentials:

```
const char* ssid      = "your_ssid";
const char* password = "your_password";
```

4. Define the IFTTT credentials:

```
const char* host = "maker.ifttt.com";
const char* privateKey = "your_key";
```

5. Variables:

```
boolean sent = false; // signifies request has been made
boolean state = 0; // holds previous trigger state
boolean event = 0; // 1-temperatureLog, 0-moistureLog
```

6. Initialize the serial port and the DHT11 sensor, and connect to the Wi-Fi hotspot:

```
void setup() {
  Serial.begin(115200); // initialize serial communication
dht.begin();
  delay(100);

  // We start by connecting to a WiFi network

  Serial.println();
  Serial.println();
  Serial.print("Connecting to ");
  Serial.println(ssid);

  WiFi.begin(ssid, password);

  while (WiFi.status() != WL_CONNECTED) {
    delay(500);
    Serial.print(".");
  }
```

```
    Serial.println("");
    Serial.println("WiFi connected");
    Serial.println("IP address: ");
    Serial.println(WiFi.localIP());
}
```

7. Read the DHT11 temperature reading and the soil moisture analog reading:

```
void loop() {
delay(5000);
    // Read temperature as Celsius (the default)
    float t = dht.readTemperature();
    // Check if any reads failed and exit early (to try again).
    if (isnan(t)) {
      Serial.println("Failed to read from DHT sensor!");
      return;
    }

    int moisture = analogRead(A0); // get soil moisture reading
```

8. Set control parameters depending on the input from the sensors:

```
    if(t < 20 && state == 0){// temperature is low
      state = 0; // turn off ventilation system
      event = 1;
      sent = false;
    }
    else if(t >= 20 && state == 1){ // temperature is high
      state = 1; // turn on ventilation system
      event = 1;
      sent = false;
    }

    if(moisture < 512 && state == 0 && sent){ // moisture low
      state = 1; // turn on pump
      event = 0;
      sent = false;
    }
    else if(moisture >= 512 && state == 1 && sent){ // moisture high
      state = 0; // turn off pump
      event = 0;
      sent = false;
    }
```

9. If the conditions have been triggered:

```
if(!sent){
    Serial.print("connecting to ");
    Serial.println(host);
    // Use WiFiClient class to create TCP connections
    WiFiClient client;
    const int httpPort = 80;
    if (!client.connect(host, httpPort)) {
      Serial.println("connection failed");
      return;
    }
```

10. Generate the URL:

```
    // We now create a URI for the request
    String url = "/trigger/";
url += event?"temperatureLog":"moistureLog";
    url += "/with/key/";
    url += privateKey;
    url += "?value1=";
    url += String(toggleState);

    Serial.print("Requesting URL: ");
    Serial.println(url);
```

11. Send the GET request and read incoming data from the server:

```
    // This will send the request to the server
    client.print(String("GET ") + url + " HTTP/1.1\r\n" +
"Host: " + host + "\r\n" +
"Connection: close\r\n\r\n");
    unsigned long timeout = millis();
    while (client.available() == 0) {
      if (millis() - timeout > 5000) {
        Serial.println(">>> Client Timeout !");
        client.stop();
        return;
      }
    }

    // Read all the lines of the reply from server and print them
to Serial
    while(client.available()){
      String line = client.readStringUntil('\r');
      Serial.print(line);
    }
```

```
        Serial.println();
        Serial.println("closing connection");
        lastTime = millis(); // save time of last trigger
    sent = true; // prevent request from being sent
      }
    }
```

Before uploading the sketches, replace `your_ssid` in the sketch with the SSID of your Wi-Fi network and `your_password` with the password of your Wi-Fi network. Upload the code and open the serial monitor to check on the status messages.

How it works...

The ESP8266 sensor hub reads the temperature measurement from the DHT11 and the soil moisture level from the soil moisture sensor. It then compares the read values to the set thresholds. The thresholds are 20 degrees Celsius for the temperature and 512 for the soil moisture reading. The ESP8266 board will check whether the readings are less than or higher than the set thresholds.

If any of the set conditions are met, a trigger is sent to the IFTTT maker channel. The IFTTT maker channel then transfers the trigger to the respective feed on Adafruit IO through the Adafruit IFTTT channel. There are two feeds on Adafruit IO, one for the temperature trigger (`temperatureLog`) and the other for the moisture trigger (`moistureLog`). If the temperature reading has met any of the set conditions, a control bit is sent to the `temperatureLog` feed. If the moisture reading meets the set conditions, a control bit is sent to the `moistureLog` feed.

If there is new data available on the Adafruit IO MQTT broker, the ESP8266 actuator hub reads it and controls the pump and relay accordingly. To do this, the ESP8266 subscribes to both the `temperatureLog` and `moistureLog` feeds. If the incoming character on the `temperatureLog` feed is 0, the relay signal pin is set to `LOW`, and if the incoming character is 1, the relay signal pin is set to `HIGH`. The same happens to the pump signal pin, depending on readings on the `moistureLog` feed.

See also

There are several issues that you may encounter when implementing M2M interactions using ESP8266 boards. The next recipe will look at some of the issues, and solutions that you can use when you encounter them.

Troubleshooting common issues with web services

In this recipe, we will discuss some of the problems you may run into and how to troubleshoot and solve them.

The board is not connecting to the Wi-Fi network

This usually happens if the Wi-Fi SSID and password provided in the code do not match those of the Wi-Fi network the ESP8266 is supposed to connect to. You can solve this by providing the correct credentials in your code.

The board is not connecting to Adafruit IO

This can happen due to the use of the wrong Adafruit IO credentials, such as your key and username. Therefore, cross-check to make sure youhave provided the correct credentials.

The board receives more than one reading or different readings from what was sent

When using the MQTT-dashboard broker (`broker.mqtt-dashboard.com`) you may sometimes receive more than one reading, or a reading that is completely different from what has been published by your other ESP8266 board. This happens when you do not use a unique topic to publish your data. That way, other devices out there publish their data to the same topic, hence your subscribed devices will receive data from other publishers. To solve this, use a unique topic.

My board is not successfully creating a hotspot

If your ESP8266 board does not create a hotspot, make sure the password you have set is at least eight characters long. That should solve the issue.

Index